I0038451

Heat Transfer: Fundamentals and Applications

Heat Transfer: Fundamentals and Applications

Naomi Watts

NY RESEARCH PRESS
New York

Published by NY Research Press
118-35 Queens Blvd., Suite 400,
Forest Hills, NY 11375, USA
www.nyresearchpress.com

Heat Transfer: Fundamentals and Applications
Naomi Watts

© 2020 NY Research Press

International Standard Book Number: 978-1-63238-807-0 (Hardback)

This book contains information obtained from authentic and highly regarded sources. All chapters are published with permission under the Creative Commons Attribution Share Alike License or equivalent. A wide variety of references are listed. Permissions and sources are indicated; for detailed attributions, please refer to the permissions page. Reasonable efforts have been made to publish reliable data and information, but the authors, editors and publisher cannot assume any responsibility for the validity of all materials or the consequences of their use.

Trademark Notice: Registered trademark of products or corporate names are used only for explanation and identification without intent to infringe.

Cataloging-in-Publication Data

Heat transfer : fundamentals and applications / Naomi Watts.
p. cm.
Includes bibliographical references and index.
ISBN 978-1-63238-807-0
1. Heat--Transmission. 2. Energy transfer. 3. Engineering. I. Watts, Naomi.
QC320 .H43 2020
621.402 2--dc23

Table of Contents

Preface

It is with great pleasure that I present this book. It has been carefully written after numerous discussions with my peers and other practitioners of the field. I would like to take this opportunity to thank my family and friends who have been extremely supporting at every step in my life.

The discipline of thermal engineering which is concerned with the generation, use, conservation and exchange of thermal energy between physical systems is referred to as heat transfer. The transfer of mass of varying chemical species to achieve heat transfer is also explored under this field. The fundamental modes of heat transfer include advection, conduction, convection and radiation. Advection is the transport mechanism of fluid that depends on its motion and momentum. Conduction is the transfer of energy between objects that are in physical contact. Convection refers to the transfer of energy between an object and its environment due to fluid motion. The transfer of energy by the emission of electromagnetic radiation is known as radiation. This book covers in detail some existent theories and innovative concepts revolving around heat transfer. Those in search of information to further their knowledge will be greatly assisted by it. Coherent flow of topics, student-friendly language and extensive use of examples make this textbook an invaluable source of knowledge.

The chapters below are organized to facilitate a comprehensive understanding of the subject:

Chapter - Heat and Thermodynamics

Heat is defined as the energy that is transferred from one body to another as the result of a difference in temperature. The important laws of thermodynamics are zeroth law, first law, second law and third law of thermodynamics. This is an introductory chapter which will introduce briefly all these significant aspects of heat and thermodynamics.

Chapter - Radiation Heat Transfer

Electromagnetic radiation generated by the thermal motion of particles in matter is referred to as radiation heat transfer. Some of its basic concepts include black body radiation, Stefan-Boltzmann law, Planck's law, etc. This chapter discusses in detail all these concepts related to radiation heat transfer.

Chapter - Understanding Convection

The heat transfer between particles due to the bulk movement of molecules within fluids such as gases and liquids is called convection. Advection, natural convection and granular convection are some of the types and mechanisms of convection. All these diverse aspects of convection have been carefully analyzed in this chapter.

Chapter - Condensation and Boiling Heat Transfer

Condensation heat transfer is the heat transfer accompanied by condensation. The heat transfer accompanied by boiling is referred to as boiling heat transfer. The different phenomena related to conduction and boiling heat transfer are film condensation, nucleate boiling, etc. These topics elaborated in this chapter will help in gaining a better perspective about condensation and boiling heat transfer.

Chapter - Mass Transfer

The net movement of mass from one location to another occurring in different processes like absorption and distillation is called mass transfer. The key concepts related to mass transfer are mass transfer coefficient, mass flux and chemical equilibrium. This chapter has been carefully written to provide an easy understanding of these facets of mass transfer.

Naomi Watts

1

Heat and Thermodynamics

Heat is defined as the energy that is transferred from one body to another as the result of a difference in temperature. The important laws of thermodynamics are zeroth law, first law, second law and third law of thermodynamics. This is an introductory chapter which will introduce briefly all these significant aspects of heat and thermodynamics.

HEAT

Heat is the energy that is transferred from one body to another as the result of a difference in temperature. If two bodies at different temperatures are brought together, energy is transferred—i.e., heat flows—from the hotter body to the colder. The effect of this transfer of energy usually, but not always, is an increase in the temperature of the colder body and a decrease in the temperature of the hotter body. A substance may absorb heat without an increase in temperature by changing from one physical state (or phase) to another, as from a solid to a liquid (melting), from a solid to a vapour (sublimation), from a liquid to a vapour (boiling), or from one solid form to another (usually called a crystalline transition). The important distinction between heat and temperature (heat being a form of energy and temperature a measure of the amount of that energy present in a body) was clarified during the 18th and 19th centuries.

Heat as a Form of Energy

Because all of the many forms of energy, including heat, can be converted into work, amounts of energy are expressed in units of work, such as joules, foot-pounds, kilowatt-hours, or calories. Exact relationships exist between the amounts of heat added to or removed from a body and the magnitude of the effects on the state of the body. The two units of heat most commonly used are the calorie and the British thermal unit (BTU). The calorie (or gram-calorie) is the amount of energy required to raise the temperature of one gram of water from 14.5° to 15.5° C; the BTU is the amount of energy required to raise the temperature of one pound of water from 63° to 64° F. One BTU is approximately 252 calories. Both definitions specify that the temperature changes are to be measured at a constant pressure of one atmosphere, because the amounts of energy involved depend in part on pressure. The calorie used in measuring the energy content of foods is the large calorie, or kilogram-calorie, equal to 1,000 gram-calories.

In general, the amount of energy required to raise a unit mass of a substance through a specified temperature interval is called the heat capacity, or the specific heat, of that substance. The quantity of energy necessary to raise the temperature of a body one degree varies depending upon the restraints imposed. If heat is added to a gas confined at constant volume, the amount of heat needed to cause a one-degree temperature rise is less than if the heat is added to the same gas free to expand (as in a cylinder fitted with a movable piston) and so do work. In the first case, all the energy goes into raising the temperature of the gas, but in the second case, the energy not only contributes to the temperature increase of the gas but also provides the energy necessary for the work done by the gas on the piston. Consequently, the specific heat of a substance depends on these conditions. The most commonly determined specific heats are the specific heat at constant volume and the specific heat at constant pressure. The heat capacities of many solid elements were shown to be closely related to their atomic weights by the French scientists Pierre-Louis Dulong and Alexis-Therese Petit in 1819. The so-called law of Dulong and Petit was useful in determining the atomic weights of certain metallic elements, but there are many exceptions to it; the deviations were later found to be explainable on the basis of quantum mechanics.

It is incorrect to speak of the heat in a body, because heat is restricted to energy being transferred. Energy stored in a body is not heat (nor is it work, as work is also energy in transit). It is customary, however, to speak of sensible and latent heat. The latent heat, also called the heat of vaporization, is the amount of energy necessary to change a liquid to a vapour at constant temperature and pressure. The energy required to melt a solid to a liquid is called the heat of fusion, and the heat of sublimation is the energy necessary to change a solid directly to a vapour, these changes also taking place under conditions of constant temperature and pressure.

Air is a mixture of gases and water vapour, and it is possible for the water present in the air to change phase; i.e., it may become liquid (rain) or solid (snow). To distinguish between the energy associated with the phase change (the latent heat) and the energy required for a temperature change, the concept of sensible heat was introduced. In a mixture of water vapour and air, the sensible heat is the energy necessary to produce a particular temperature change excluding any energy required for a phase change.

HEAT TRANSFER

Heat transfer describes the flow of heat (thermal energy) due to temperature differences and the subsequent temperature distribution and changes.

The study of transport phenomena concerns the exchange of momentum, energy and mass in the form of conduction, convection, and radiation. These processes can be described via mathematical formulas.

The fundamentals for these formulas are found in the laws for conservation of momentum, energy and mass in combination with constitutive laws, relations that describe not only the conservation but also the flux of quantities involved in these phenomena. For that purpose, differential equations are used to describe the mentioned laws and constitutive relations in the best way possible. Solving these equations is an effective way to investigate systems and predict their behavior.

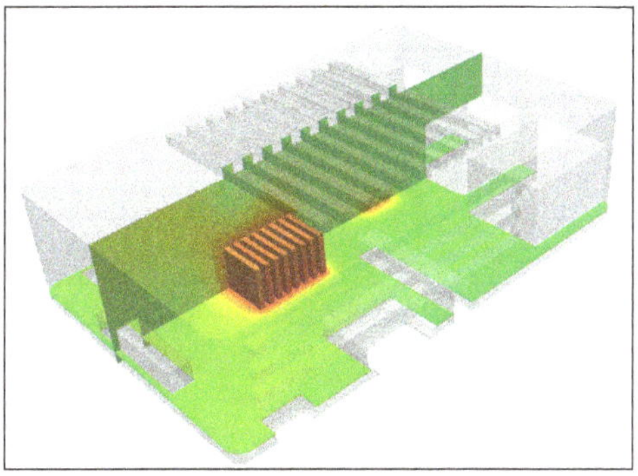

Heat Sink cooling with Sim Scale.

Without external help, heat will always flow from hot objects to cold ones which is a direct consequence of the second law of thermodynamics.

We call that heat flow. In the early nineteenth century, scientists believed that all bodies contained an invisible fluid called caloric (a massless fluid thought to flow from hot to cold objects). Caloric was assigned properties, some of which proved to be inconsistent with nature (for instance it had weight and it could not be created nor destroyed). But its most important feature was that it was able to flow from hot bodies into cold ones. That was a very useful way to think about heat.

Thompson and Joule showed that this theory of the caloric was wrong. Heat is not a substance as supposed, but a motion at the molecular level (so called kinetic theory). A good example is rubbing our hands against each other. Both hands get warmer, even though initially they were at the same cooler temperatures. Now if the cause of the heat was a fluid, then it would have flowed from a (hotter) body with more energy to another with less energy (colder). Instead, the hands are heated because the kinetic energy of motion (rubbing) has been converted to heat in a process called “friction”.

The flow of heat is happening all the time from any physical entity to objects surrounding it. Heat flows constantly from your body to the air surrounding you. Small buoyancy-driven (or convective) motion of the air will continue in a room because the walls can never be perfectly isothermal as in theory. The only domain free from heat flow would have to be isothermal and completely isolated from any other system allowing heat transfer.

The cooling of the sun are the primary processes that we experience naturally. Other processes are the conductive cooling of Earth's center and the radiative cooling of other stars.

Phenomenology

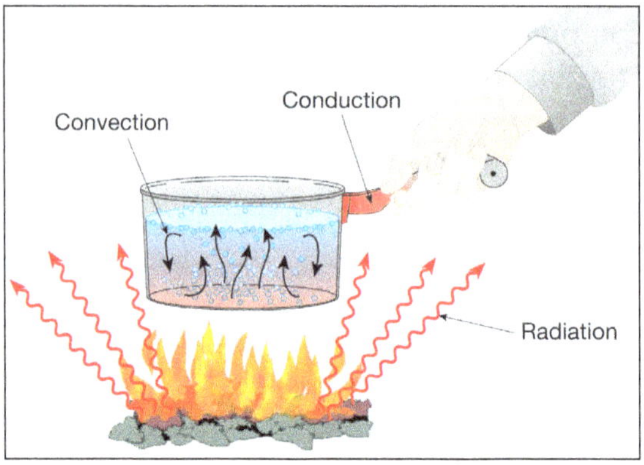

Conduction, Convection and Radiation explained in one picture.

Conduction

Fourier's law: Joseph Fourier published his book "Théorie Analytique de la Chaleur" in 1822.

Joseph Fourier - French mathematician and physicist.

He formulated a very complete theory of heat conduction. He stated the empirical law that bears his name: the heat flux (indicated with $q -$ a heat rate per unit area and can be expressed as $\frac{Q}{A}$), $q(W/m^2)$, resulting from thermal conduction is proportional to the magnitude of the temperature gradient. If we name the constant of proportionality, k, that means

$$q = -k\frac{dT}{dx}$$

The constant, k, is called the thermal conductivity with the dimensions $\frac{W}{m*K}$, or $\frac{J}{m*s*K}$.

Please keep in mind that the heat flux is a vector quantity. Equation $q=-k\frac{dT}{dx}$ tells us that, if temperature decreases with x, q will be positive - it will flow in positive x -direction. If T increases with x, q will be negative; it will flow in negative x-direction. In either case, q will flow from higher temperatures to lower temperatures as already mentioned in the introductory paragraph. Equation $q=-k\frac{dT}{dx}$ is the one-dimensional formulation of Fourier's law. The three-dimensional equivalent form is:

$$\vec{q}=-k\nabla T$$

Where ∇ indicates the so called gradient.

In one-dimensional heat conduction problems, there is no problem to tell in which way the heat flows. For that reason, it is often convenient to write Fourier's law in simple scalar form:

$$q=k\frac{\Delta T}{L}$$

Where L is the thickness in the direction of heat flow and q and ΔT are both written as positive quantities. We just have to keep in mind that q always flows from high to low temperature.

The thermal conductivity of gases can be understood with the imagination of molecules. These molecules move through thermal movement from one position to another position as can be seen in the picture below.

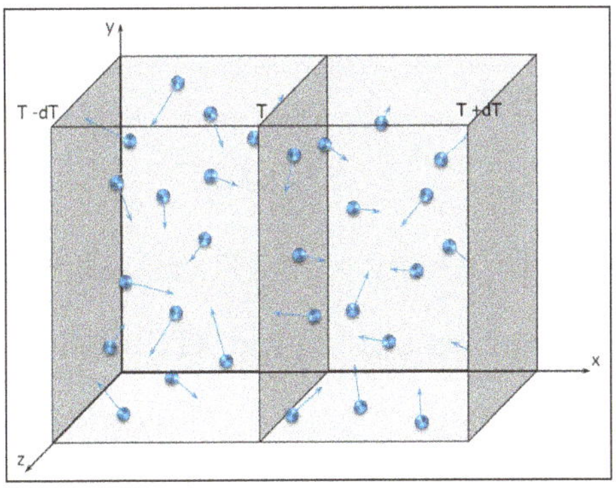

Thermal conductivity of gas.

The internal energy of the molecules is transferred by impact with other molecules. Areas with low temperature will be occupied by molecules of high temperature and areas with high temperature will be occupied by molecules of lower temperature. The thermal conductivity can be explained with this imagination and be derived with the kinetic theory of gases:

$$T = \frac{2}{3}\frac{K}{Nk_B}$$

Which states that "the average molecular kinetic energy is proportional to the ideal gas law's absolute temperature". For an ideal gas the thermal conductivity is independent from the pressure and increases with the root of the temperature.

This theory is pretty hard to understand for objects other than metals. And for fluids it is even more difficult because there is no simple theory. In nonmetallic components, heat transfers via lattice vibrations (Phonon). The thermal conductivity transferred by phonons also exists in metals but surpassed by the conductivity of the electron gas.

The low thermal conductivity of insulating materials like polystyrene or glass wool is based on the principal of the low thermal conductivity of air (or any other gases).

Table: Thermal conductivity of different materials.

Material	Thermal conductivity W/(m.K)
Oxygen	0.023
Steam	0.0248
Polystyrene	0.032-0.050
Water	0.5562
Glass	0.76
Concrete	2.1
Steel high-alloyed	15
Steel unalloyed	48-58
Iron	80.2
Copper pure	401
Diamond	2300

Analogous

- Heat Transfer: Heat flux density $\propto$ T (Thermal conductivity).
- Diffusion: Partial current density $\propto$ grad x (Diffusion coefficient).
- Electric lead: Current density $\propto$ grad U_{el} (Electric conductivity).

Radiation

Radiation describes the phenomenon of transmission of energy from one body to another by propagation through a medium. All bodies constantly emit energy by electromagnetic radiation. The intensity of such energy flux depends not only on the temperature of the body but also on the surface characteristics. If you sit in front of a camp fire, most of the heat that reaches you is radiant energy. Very often, emission of energy, or radiant heat transfer, from cooler bodies can be neglected in comparison to convection and conduction. Heat transfer processes happening at high temperature, or with conduction or convection suppressed by evacuated insulation, involve a significant fraction of radiation in general.

The electromagnetic (EM) spectrum: This spectrum is the range of all types of electromagnetic radiation. Simply put, radiation is energy travelling and spreading out like photons being emitted by a lamp or radio waves. Other well known types of electromagnetic radiation are X-Rays, gamma-rays, microwaves, infrared light etc.

Electromagnetic radiation can be seen as a stream of photons, each traveling in a wave-like pattern, moving at the speed of light and carrying energy. The difference between the different forms in the electromagnetic spectrum is the energy of the photons. It is important to keep in mind that if we talk about the energy of a photon, the behavior can either be that of a wave or that of a particle which we call the "wave-particle duality" of light.

Each quantum of radiant energy has a wavelength λ and a frequency, v, associated with it. The relation between energy, wavelength, λ and frequency, v, can be written as wavelength equals the speed of light divided by the frequency, or

$$\lambda = \frac{c}{v}$$

and energy equals Planck's constant times the frequency, or

$$E = h * v$$

Where h is Planck's constant ($6{,}626070040 * 10^{-34} Js$).

The table below shows various forms over a range of wavelengths. Thermal radiation is from 0.1-1000 μm.

Table: Electromagnetic wave spectrum.

Characterization	Wavelength
Gamma rays	0.3 100 pm
X-rays	0.01-30 nm
Ultraviolet light	3-400 nm
Visible light	0.4-0.7 µm
Near infrared radiation	0.7-30 µm
Far infrared radiation	30-1000 µm
Microwaves	10-300 mm
Shortwave radio & TV	300 mm-100 m

If radiation meets a body or a fluid it will be:

- Reflected,
- Transmitted,
- Absorbed.

A Body itself can also Emit Radiation

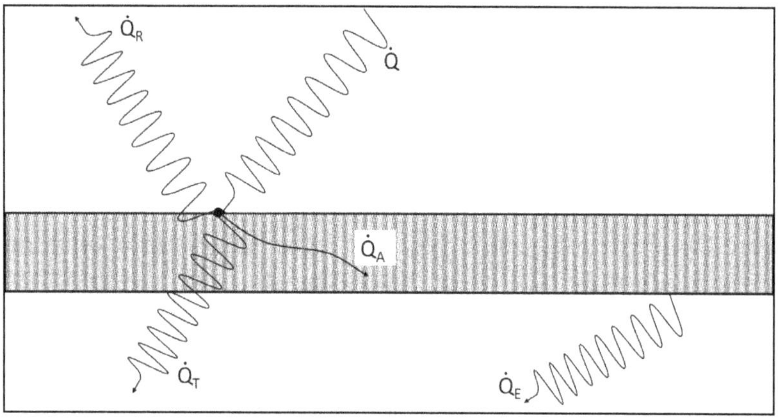

Radiation of a Body.

$$\dot{Q} = \dot{Q}A + \dot{Q}T + \dot{Q}R$$

$$1 = \frac{\dot{Q}_A}{\dot{Q}} + \frac{\dot{Q}_T}{\dot{Q}} + \frac{\dot{Q}_R}{\dot{Q}}$$

$$1 = \alpha^S + \tau^S + \rho^S$$

Where,

α^S: Absorptance

τ^S : Transmittance

ρ^S : Reflectance

- Black Body: $\alpha^S = 1 \quad \rho^S = 0 \quad \tau^S = 0.$
- Gray Body: α^S, ρ^S and τ^S uniform for all wavelengths.
- White Body: $\alpha^S = 0 \quad \rho^S = 1 \quad \tau^S = 0.$
- Opaque Body: $\alpha^S + 0 \quad \rho^S = 1 \quad \tau^S = 0.$
- TransparentBody: $\alpha^S = 0 \quad \rho^S = 0 \quad \tau^S = 1.$

Black Body

"Blackbody radiation" refers to an object or system in thermodynamic equilibrium which absorbs all incoming radiation and emits energy of a characteristic, temperature dependent spectrum. This behavior is specific of this radiating system only and is not dependent on the type of radiation which is incident upon it.

Stefan-boltzmann Law

The thermal energy radiated by a blackbody radiator per second per unit area is proportional to the fourth power of the absolute temperature and is given by:

$$\frac{P}{A} = \sigma T^4$$

Where σ is the Stefan-Boltzmann constant which can be derived from other constants of nature:

$$\sigma = \frac{2\pi^5 k^4}{15c^2 h^3} = 5.670373 * 10^{-8} Wm^{-2} K^{-4}$$

For hot objects other than ideal radiators, the law is expressed in the form:

$$\frac{P}{A} = e\sigma T^4$$

where e is the emissivity of the object ($e = 1$ for ideal radiator). If the hot object is radiating energy to its colder surroundings at temperature T_c, the net radiation loss rate takes the form:

$$P = e\sigma A(T^4 - T_c^4)$$

Due to the fourth power of the temperatures in the governing equation, radiation becomes a very complex, high-level nonlinear phenomenon.

Convection

Consider a convective cooling situation. Cold gas flows past a warm body as shown in the figure below:

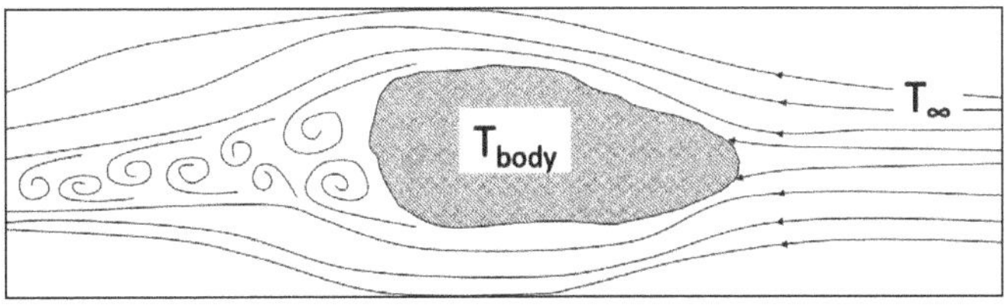

Convective cooling of a heated body.

The fluid forms immediately adjacent to the body a thin slowed-down region called a boundary layer. Heat is conducted into this layer, which vanishes and mixes into the stream. We call this process of carrying heat away by a moving fluid convection.

Sir Isaac Newton - English mathematician, astronomer and physicist.

Isaac Newton (1701) considered the convective process and suggested a simple formula for the cooling:

$$\frac{dT_{body}}{dt} \propto T_{body} - T_{\infty}$$

Where T_{∞} is the temperature of the oncoming fluid. This expression proposes that energy is flowing away from the body.

The steady-state form of Newton's Law of cooling defining free convection is described by the following formula:

$$Q = h(T_{body} - T_{\infty})$$

Where h is the heat transfer coefficient. This coefficient can be denoted with a bar $\bar{h}$ which indicates the average over the surface of the body. h without a bar it denotes the "local" values of the coefficient.

Depending on how the fluid motion is initiated, we can classify convection as natural (free) or forced convection. Natural convection is caused for instance by buoyancy effects (warm fluid rises and cold fluid falls). In the other case, forced convection causes the fluid to move by external means such as a fan, wind, coolant, pump, suction devices, etc.

Forced convection: The movement of a solid component into a fluid can also be considered as forced convection. Natural convection can create a noticeable temperature difference in a house or flat. We recognize this because certain parts of the house are warmer than others. Forced convection creates a more uniform temperature distribution and therefore comfortable feeling throughout the entire home.

This reduces cold spots in the house, reducing the need to crank the thermostat to a higher temperature.

CFD Heat Analysis – Structural Heat Transfer

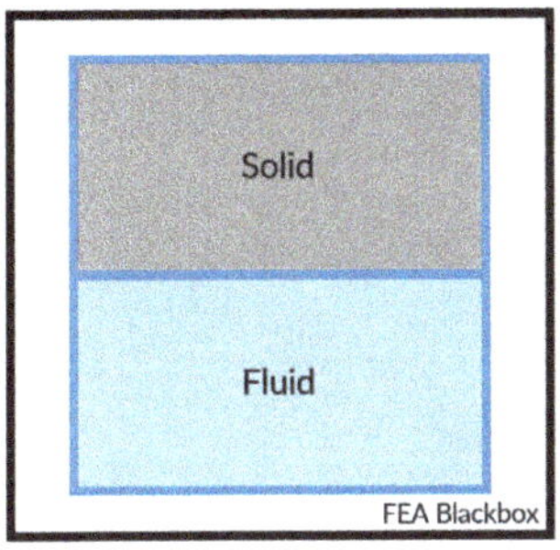

Fluid-Solid Analysis

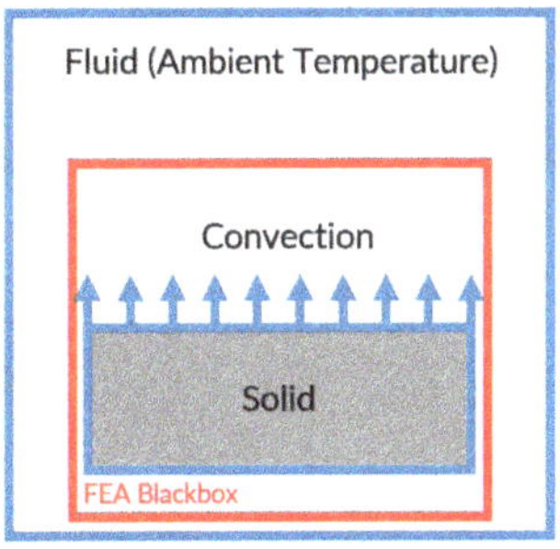

Heat Transfer Analysis (Structural)

Structural heat transfer analysis compared to fluid-solid analysis.

Structural Heat Transfer Analysis is used when:

- The fluid temperature can be assumed to be homogeneous around the solid part.
- Investigating the behavior of structural components only under heating.
- Investigating stress and deformation by the part caused by the heat load (thermal stress analysis).

Coupled Heat Transfer Analysis (Fluid-Solid) used when:

- The fluid distribution around the solid needs to be studied.
- Investigating the influence of the object.
- Investigating natural cooling.

Heat Transfer Analysis – Linear Static Analysis

Table: Heat Transfer Analysis compared to Structural Analysis.

Category	Structural Analysis (linear static)	Heat Transfer Analysis (steady state)
Material properties	Young's modulus(E)	Thermal conductivity(k)
Laws	Hook's law $\sigma = E \cdot \frac{du}{dx}$	Fourier law $q = -k \cdot \frac{dT}{dx}$
Degree of Freedom (DOF)	Displacement (u)	Temperature (T)

Gradient of DOF	Stain ϵ Stress σ	Temperature gradient (∇T)
Similarities	Axial force per unit length: Q Cross-sectional area: A Young's modulus: E	Internal heat generation per unit length: Q Cross-sectional area: A Thermal conductivity: k

Applications of Thermal Analysis

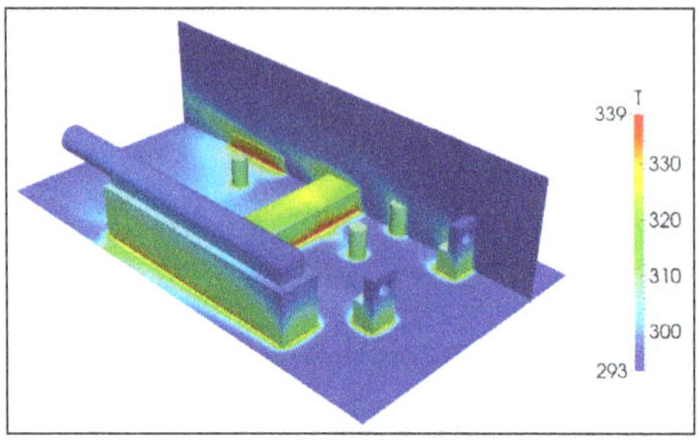

Printed Circuit Board - simulated with SimScale.

Thermal - Structural Analysis

Heat Transfer takes the energy balance of the studied systems into account. When investigating thermomechanical components, structural deformations, caused by the effects of thermal loads on solids can also be included. Simulating the stress response to thermal loads and failure is essential for many industrial applications. An example for an application is a thermal stress analysis of a Printed Circuit Board.

Conjugate Heat Transfer

Conjugate Heat Transfer (CHT) simulations analyze the coupled heat transfer in fluids and solids. The prediction of the fluid flow while simultaneously analyzing the heat transfer that takes place within the fluid/solid boundary is an important feature of CHT simulations. One of the areas in which it can be used is for electronics cooling.

Conduction

In theory, heat passes from a hot to a cold object. Conduction is the heat transfer from a hot to a cold object, that are in direct contact to each other. The thermal conductivity of the different objects decides how much heat in which time is being transferred. Examples include CFL light bulbs.

Convection

Convective Heat Transfer is the transfer of heat between two areas without physical contact. Convective currents occur when molecules absorb heat and start moving. As you can imagine, these effects are difficult to predict which is why high computing power is needed to obtain reliable results from a simulation. One application is the cooling of a Raspberry pi mother board.

Radiation

Electromagnetic waves are the source of heat transfer through radiation. They usually play a role at high temperatures. The amount of heat that is emitted via radiation depends on the surface type of the material. A general rule is that the more surface there is, the higher the radiation is. An example application where simulation of radiation is used, is the simulation of laser beam welding.

THERMODYNAMICS

Thermodynamics is the science of the relationship between heat, work, temperature, and energy. In broad terms, thermodynamics deals with the transfer of energy from one place to another and from one form to another. The key concept is that heat is a form of energy corresponding to a definite amount of mechanical work.

Heat was not formally recognized as a form of energy until about 1798, when Count Rumford (Sir Benjamin Thompson), a British military engineer, noticed that limitless amounts of heat could be generated in the boring of cannon barrels and that the amount of heat generated is proportional to the work done in turning a blunt boring tool. Rumford's observation of the proportionality between heat generated and work done lies at the foundation of thermodynamics. Another pioneer was the French military engineer Sadi Carnot, who introduced the concept of the heat-engine cycle and the principle of reversibility in 1824. Carnot's work concerned the limitations on the maximum amount of work that can be obtained from a steam engine operating with a high-temperature heat transfer as its driving force. Later that century, these ideas were developed by Rudolf Clausius, a German mathematician and physicist, into the first and second laws of thermodynamics, respectively.

The most important laws of thermodynamics are:

- The zeroth law of thermodynamics: When two systems are each in thermal equilibrium with a third system, the first two systems are in thermal equilibrium with each other. This property makes it meaningful to use thermometers as the "third system" and to define a temperature scale.
- The first law of thermodynamics, or the law of conservation of energy: The change in a system's internal energy is equal to the difference between heat added to the system from its surroundings and work done by the system on its surroundings.
- The second law of thermodynamics: Heat does not flow spontaneously from a colder region to a hotter region, or, equivalently, heat at a given temperature

cannot be converted entirely into work. Consequently, the entropy of a closed system, or heat energy per unit temperature, increases over time toward some maximum value. Thus, all closed systems tend toward an equilibrium state in which entropy is at a maximum and no energy is available to do useful work. This asymmetry between forward and backward processes gives rise to what is known as the "arrow of time."

- The third law of thermodynamics: The entropy of a perfect crystal of an element in its most stable form tends to zero as the temperature approaches absolute zero. This allows an absolute scale for entropy to be established that, from a statistical point of view, determines the degree of randomness or disorder in a system.

Although thermodynamics developed rapidly during the 19th century in response to the need to optimize the performance of steam engines, the sweeping generality of the laws of thermodynamics makes them applicable to all physical and biological systems. In particular, the laws of thermodynamics give a complete description of all changes in the energy state of any system and its ability to perform useful work on its surroundings.

Such concerns are the focus of the branch of thermodynamics known as statistical thermodynamics, or statistical mechanics, which expresses macroscopic thermodynamic properties in terms of the behaviour of individual particles and their interactions. It has its roots in the latter part of the 19th century, when atomic and molecular theories of matter began to be generally accepted.

Fundamental Concepts

Thermodynamic States

The application of thermodynamic principles begins by defining a system that is in some sense distinct from its surroundings. For example, the system could be a sample of gas inside a cylinder with a movable piston, an entire steam engine, a marathon runner, the planet Earth, a neutron star, a black hole, or even the entire universe. In general, systems are free to exchange heat, work, and other forms of energy with their surroundings.

A system's condition at any given time is called its thermodynamic state. For a gas in a cylinder with a movable piston, the state of the system is identified by the temperature, pressure, and volume of the gas. These properties are characteristic parameters that have definite values at each state and are independent of the way in which the system arrived at that state. In other words, any change in value of a property depends only on the initial and final states of the system, not on the path followed by the system from one state to another. Such properties are called state functions. In contrast, the work done as the piston moves and the gas expands and the heat the gas absorbs from its surroundings depend on the detailed way in which the expansion occurs.

The behaviour of a complex thermodynamic system, such as Earth's atmosphere, can be understood by first applying the principles of states and properties to its component parts—in this case, water, water vapour, and the various gases making up the atmosphere. By isolating samples of material whose states and properties can be controlled and manipulated, properties and their interrelations can be studied as the system changes from state to state.

Thermodynamic Equilibrium

A particularly important concept is thermodynamic equilibrium, in which there is no tendency for the state of a system to change spontaneously. For example, the gas in a cylinder with a movable piston will be at equilibrium if the temperature and pressure inside are uniform and if the restraining force on the piston is just sufficient to keep it from moving. The system can then be made to change to a new state only by an externally imposed change in one of the state functions, such as the temperature by adding heat or the volume by moving the piston. A sequence of one or more such steps connecting different states of the system is called a process. In general, a system is not in equilibrium as it adjusts to an abrupt change in its environment. For example, when a balloon bursts, the compressed gas inside is suddenly far from equilibrium, and it rapidly expands until it reaches a new equilibrium state. However, the same final state could be achieved by placing the same compressed gas in a cylinder with a movable piston and applying a sequence of many small increments in volume (and temperature), with the system being given time to come to equilibrium after each small increment. Such a process is said to be reversible because the system is at (or near) equilibrium at each step along its path, and the direction of change could be reversed at any point. This example illustrates how two different paths can connect the same initial and final states. The first is irreversible (the balloon bursts), and the second is reversible. The concept of reversible processes is something like motion without friction in mechanics. It represents an idealized limiting case that is very useful in discussing the properties of real systems. Many of the results of thermodynamics are derived from the properties of reversible processes.

Temperature

The concept of temperature is fundamental to any discussion of thermodynamics, but its precise definition is not a simple matter. For example, a steel rod feels colder than a wooden rod at room temperature simply because steel is better at conducting heat away from the skin. It is therefore necessary to have an objective way of measuring temperature. In general, when two objects are brought into thermal contact, heat will flow between them until they come into equilibrium with each other. When the flow of heat stops, they are said to be at the same temperature. The zeroth law of thermodynamics formalizes this by asserting that if an object A is in simultaneous thermal equilibrium with two other objects B and C, then B and C will be in thermal equilibrium with each other if brought into thermal contact. Object A can then play the role of a thermometer through some change in its physical properties with temperature, such as its volume or its electrical resistance.

With the definition of equality of temperature in hand, it is possible to establish a temperature scale by assigning numerical values to certain easily reproducible fixed points. For example, in the Celsius (°C) temperature scale, the freezing point of pure water is arbitrarily assigned a temperature of 0 °C and the boiling point of water the value of 100 °C. In the Fahrenheit (°F) temperature scale, these same two points are assigned the values 32 °F and 212 °F, respectively. There are absolute temperature scales related to the second law of thermodynamics. The absolute scale related to the Celsius scale is called the Kelvin (K) scale, and that related to the Fahrenheit scale is called the Rankine (°R) scale. These scales are related by the equations K = °C + 273.15, °R = °F + 459.67, and °R = 1.8 K. Zero in both the Kelvin and Rankine scales is at absolute zero.

Work and Energy

Energy has a precise meaning in physics that does not always correspond to everyday language, and yet a precise definition is somewhat elusive. For example, a man pushing on a car may feel that he is doing a lot of work, but no work is actually done unless the car moves. The work done is then the product of the force applied by the man multiplied by the distance through which the car moves. If there is no friction and the surface is level, then the car, once set in motion, will continue rolling indefinitely with constant speed. The rolling car has something that a stationary car does not have—it has kinetic energy of motion equal to the work required to achieve that state of motion. The introduction of the concept of energy in this way is of great value in mechanics because, in the absence of friction, energy is never lost from the system, although it can be converted from one form to another. For example, if a coasting car comes to a hill, it will roll some distance up the hill before coming to a temporary stop. At that moment its kinetic energy of motion has been converted into its potential energy of position, which is equal to the work required to lift the car through the same vertical distance. After coming to a stop, the car will then begin rolling back down the hill until it has completely recovered its kinetic energy of motion at the bottom. In the absence of friction, such systems are said to be conservative because at any given moment the total amount of energy (kinetic plus potential) remains equal to the initial work done to set the system in motion.

As the science of physics expanded to cover an ever-wider range of phenomena, it became necessary to include additional forms of energy in order to keep the total amount of energy constant for all closed systems (or to account for changes in total energy for open systems). For example, if work is done to accelerate charged particles, then some of the resultant energy will be stored in the form of electromagnetic fields and carried away from the system as radiation. In turn the electromagnetic energy can be picked up by a remote receiver (antenna) and converted back into an equivalent amount of work. With his theory of special relativity, Albert Einstein realized that energy (E) can also be stored as mass (m) and converted back into energy, as expressed by his famous equation $E = mc^2$, where c is the velocity of light. All of these systems are said to be conservative in the sense that energy can be freely converted from one form to another without limit. Each fundamental advance of physics into new realms has involved a similar extension to the list of the different forms of energy. In addition to preserving the first law of thermodynamics, also called the law of conservation of energy, each form of energy can be related back to an equivalent amount of work required to set the system into motion.

Thermodynamics encompasses all of these forms of energy, with the further addition of heat to the list of different kinds of energy. However, heat is fundamentally different from the others in that the conversion of work (or other forms of energy) into heat is not completely reversible, even in principle. In the example of the rolling car, some of the work done to set the car in motion is inevitably lost as heat due to friction, and the car eventually comes to a stop on a level surface. Even if all the generated heat were collected and stored in some fashion, it could never be converted entirely back into mechanical energy of motion. This fundamental limitation is expressed quantitatively by the second law of thermodynamics.

The role of friction in degrading the energy of mechanical systems may seem simple and obvious, but the quantitative connection between heat and work, as first discovered by Count Rumford, played a key role in understanding the operation of steam engines in the 19th century and similarly for all energy-conversion processes today.

Total Internal Energy

Although classical thermodynamics deals exclusively with the macroscopic properties of materials—such as temperature, pressure, and volume—thermal energy from the addition of heat can be understood at the microscopic level as an increase in the kinetic energy of motion of the molecules making up a substance. For example, gas molecules have translational kinetic energy that is proportional to the temperature of the gas: the molecules can rotate about their centre of mass, and the constituent atoms can vibrate with respect to each other (like masses connected by springs). Additionally, chemical energy is stored in the bonds holding the molecules together, and weaker long-range interactions between the molecules involve yet more energy. The sum total of all these forms of energy constitutes the total internal energy of the substance in a given thermodynamic state. The total energy of a system includes its internal energy plus any other forms of energy, such as kinetic energy due to motion of the system as a whole (e.g., water flowing through a pipe) and gravitational potential energy due to its elevation.

Thermodynamic Properties and Relations

In order to carry through a program of finding the changes in the various thermodynamic functions that accompany reactions—such as entropy, enthalpy, and free energy—it is often useful to know these quantities separately for each of the materials entering into the reaction. For example, if the entropies are known separately for the reactants and products, then the entropy change for the reaction is just the difference,

$$\Delta S_{reaction} = S_{products} - S_{reactants}$$

and similarly for the other thermodynamic functions. Furthermore, if the entropy change for a reaction is known under one set of conditions of temperature and pressure, it can be found under other sets of conditions by including the variation of entropy for the reactants and products with temperature or pressure as part of the overall process. For these reasons, scientists and engineers have developed extensive tables of thermodynamic properties for many common substances, together with their rates of change with state variables such as temperature and pressure.

The science of thermodynamics provides a rich variety of formulas and techniques that allow the maximum possible amount of information to be extracted from a limited number of laboratory measurements of the properties of materials. However, as the thermodynamic state of a system depends on several variables—such as temperature, pressure, and volume—in practice it is necessary first to decide how many of these are independent and then to specify what variables are allowed to change while others are held constant. For this reason, the mathematical language of partial differential equations is indispensable to the further elucidation of the subject of thermodynamics.

Of especially critical importance in the application of thermodynamics are the amounts of work required to make substances expand or contract and the amounts of heat required to change the temperature of substances. The first is determined by the equation of state of the substance and the second by its heat capacity. Once these physical properties have been fully characterized, they can be used to calculate other thermodynamic properties, such as the free energy of the substance under various conditions of temperature and pressure.

In what follows, it will often be necessary to consider infinitesimal changes in the parameters specifying the state of a system. The first law of thermodynamics then assumes the differential form $dU = d'Q - d'W$. Because U is a state function, the infinitesimal quantity dU must be an exact differential, which means that its definite integral depends only on the initial and final states of the system. In contrast, the quantities $d'Q$ and $d'W$ are not exact differentials, because their integrals can be evaluated only if the path connecting the initial and final states is specified. The examples to follow will illustrate these rather abstract concepts.

Work of Expansion and Contraction

The first task in carrying out the above program is to calculate the amount of work done by a single pure substance when it expands at constant temperature. Unlike the case of a chemical reaction, where the volume can change at constant temperature and pressure because of the liberation of gas, the volume of a single pure substance placed in a cylinder cannot change unless either the pressure or the temperature changes. To calculate the work, suppose that a piston moves by an infinitesimal amount dx. Because pressure is force per unit area, the total restraining force exerted by the piston on the gas is PA, where A is the cross-sectional area of the piston. Thus, the incremental amount of work done is $d'W = PA\,dx$.

However, A dx can also be identified as the incremental change in the volume (dV) swept out by the head of the piston as it moves. The result is the basic equation $d'W = P\,dV$ for the incremental work done by a gas when it expands. For a finite change from an initial volume V_i to a final volume V_f, the total work done is given by the integral

$$\mathrm{w} = \int_{v_i}^{v_f} P dv.$$

Because P in general changes as the volume V changes, this integral cannot be calculated until P is specified as a function of V; in other words, the path for the process must be specified. This gives precise meaning to the concept that dW is not an exact differential.

Equations of State

The equation of state for a substance provides the additional information required to calculate the amount of work that the substance does in making a transition from one equilibrium state to another along some specified path. The equation of state is expressed as a functional relationship connecting the various parameters needed to specify the state of the system. The basic concepts apply to all thermodynamic systems, but here, in order to make the discussion specific, a simple gas inside a cylinder with a movable piston will be considered. The equation of state then takes the form of an equation relating P, V, and T, such that if any two are specified, the third is determined. In the limit of low pressures and high temperatures, where the molecules of the gas move almost independently of one another, all gases obey an equation of state known as the ideal gas law: $PV = nRT$, where n is the number of moles of the gas and R is the universal gas constant, 8.3145 joules per K. In the International System of Units, energy is measured in joules, volume in cubic metres (m^3), force in newtons (N), and pressure in pascals (Pa), where 1 Pa = 1 N/m^2. A force of one newton moving through a distance of one metre does one joule of work. Thus, both the products PV and RT have the dimensions of work (energy). *A P-V* diagram would show the equation of state in graphical form for several different temperatures.

To illustrate the path-dependence of the work done, consider three processes connecting the same initial and final states. The temperature is the same for both states, but, in going from state i to state *f*, the gas expands from V_i to V_f (doing work), and the pressure falls from P_i to P_f. According to the definition of the integral in equation $w = \int_{v_i}^{v_f} Pdv$, the work done is the area under the curve (or straight line) for each of the three processes. For processes I and III the areas are rectangles, and so the work done is,

$$W_I = P_i(V_f - V_i)$$

and

$$W_{III} = P_f(V_f - V_i),$$

respectively. Process II is more complicated because *P* changes continuously as *V* changes. However, *T* remains constant, and so one can use the equation of state to substitute $P = nRT/V$ in equation $w = \int_{v_i}^{v_f} Pdv$. to obtain,

$$W_{II} = \int_{v_i}^{v_f} \frac{nRT}{V} dV = nRT \, In \frac{V_f}{V_i}$$

or, because,

$$P_i V_i = nRT = PfVf$$

for an (ideal gas) isothermal process,

$$W_{II} = P_i V_i \, In \frac{Vf}{Vi} = P_f V_f In \frac{Vf}{Vi}.$$

WII is thus the work done in the reversible isothermal expansion of an ideal gas. The amount of work is clearly different in each of the three cases. For a cyclic process the net work done equals the area enclosed by the complete cycle.

Heat Capacity and Specific Heat

As shown originally by Count Rumford, there is an equivalence between heat (measured in calories) and mechanical work (measured in joules) with a definite conversion factor between the two. The conversion factor, known as the mechanical equivalent of heat, is 1 calorie = 4.184 joules. (There are several slightly different definitions in use for the calorie. The calorie used by nutritionists is actually a kilocalorie.) In order to have a consistent set of units, both heat and work will be expressed in the same units of joules.

The amount of heat that a substance absorbs is connected to its temperature change via its molar specific heat c, defined to be the amount of heat required to change the temperature of 1 mole of the substance by 1 K. In other words, c is the constant of proportionality relating the heat absorbed *(d'Q)* to the temperature change *(dT)* according to $d'Q = nc\, dT$, where n is the number of moles. For example, it takes approximately 1 calorie of heat to increase the temperature of 1 gram of water by 1 K. Since there are 18 grams of water in 1 mole, the molar heat capacity of water is 18 calories per K, or about 75 joules per K. The total heat capacity *C* for n moles is defined by $C = nc$.

However, since $d'Q$ is not an exact differential, the heat absorbed is path-dependent and the path must be specified, especially for gases where the thermal expansion is significant. Two common ways of specifying the path are either the constant-pressure path or the constant-volume path. The two different kinds of specific heat are called c_P and c_V respectively, where the subscript denotes the quantity that is being held constant. It should not be surprising that c_P is always greater than c_V, because the substance must do work against the surrounding atmosphere as it expands upon heating at constant pressure but not at constant volume. In fact, this difference was used by the 19th-century German physicist Julius Robert von Mayer to estimate the mechanical equivalent of heat.

Heat Capacity and Internal Energy

The goal in defining heat capacity is to relate changes in the internal energy to measured changes in the variables that characterize the states of the system. For a system consisting of a single pure substance, the only kind of work it can do is atmospheric work, and so the first law reduces to,

$$dU = d'Q - P\,dV.$$

Suppose now that *U* is regarded as being a function *U(T, V)* of the independent pair of variables *T* and *V*. The differential quantity *dU* can always be expanded in terms of its partial derivatives according to,

$$dU = \left(\frac{\partial U}{\partial T}\right)_V dT + \left(\frac{\partial U}{\partial T}\right)_T dV$$

where the subscripts denote the quantity being held constant when calculating derivatives. Substituting this equation into $dU = d'Q - P\,dV$ then yields the general expression,

$$d'Q = \left(\frac{\partial U}{\partial T}\right)_V dT + \left[P + \left(\frac{\partial U}{\partial T}\right)_T\right] dV$$

for the path-dependent heat. The path can now be specified in terms of the independent variables *T* and *V*. For a temperature change at constant volume, $dV = 0$ and, by definition of heat capacity,

$$d'Q_V = C_V\,dT.$$

The above equation then gives immediately,

$$C_V = \left(\frac{\partial U}{\partial T}\right)_V$$

for the heat capacity at constant volume, showing that the change in internal energy at constant volume is due entirely to the heat absorbed.

To find a corresponding expression for *CP*, one need only change the independent variables to *T* and *P* and substitute the expansion,

$$dV = \left(\frac{\partial V}{\partial T}\right)_V dT + \left[P\left(\frac{\partial V}{\partial T}\right)_T\right]_T dP$$

for *dV* in equation (dU = d'Q – P dV) and correspondingly for *dU* to obtain,

$$d'Q = \left[\left(\frac{\partial U}{\partial T}\right)_P + P\left(\frac{\partial V}{\partial T}\right)_P\right] dT + \left[\left(\frac{\partial U}{\partial P}\right)_T + P\left(\frac{\partial V}{\partial P}\right)_T\right] dP$$

For a temperature change at constant pressure, $dP = 0$, and, by definition of heat capacity, $d'Q = C_p\, dT$, resulting in,

$$C_P = C_V + \left[P + \left(\frac{\partial U}{\partial V}\right)_T\right]\left(\frac{\partial V}{\partial T}\right)_P.$$

The two additional terms beyond C_V have a direct physical meaning. The term,

$$P\left(\frac{\partial V}{\partial T}\right)_P$$

represents the additional atmospheric work that the system does as it undergoes thermal expansion at constant pressure, and the second term involving

$$\left(\frac{\partial U}{\partial V}\right)_T$$

represents the internal work that must be done to pull the system apart against the forces of attraction between the molecules of the substance (internal stickiness). Because there is no internal stickiness for an ideal gas, this term is zero, and, from the ideal gas law, the remaining partial derivative is,

$$P\left(\frac{\partial V}{\partial T}\right)_P = nR.$$

With these substitutions the equation for CP becomes simply,

$$C_P = C_V + nR$$

or

$$c_P = c_V + R$$

for the molar specific heats. For example, for a monatomic ideal gas (such as helium), $c_V = 3R/2$ and $c_P = 5R/2$ to a good approximation. $c_V T$ represents the amount of translational kinetic energy possessed by the atoms of an ideal gas as they bounce around randomly inside their container. Diatomic molecules (such as oxygen) and polyatomic molecules (such as water) have additional rotational motions that also store thermal energy in their kinetic energy of rotation. Each additional degree of freedom contributes an additional amount R to c_V. Because diatomic molecules can rotate about two axes and polyatomic molecules can rotate about three axes, the values of c_V increase to 5R/2 and 3R respectively, and c_P correspondingly increases to 7R/2 and 4R. (c_V and c_P increase still further at high temperatures because of vibrational degrees of freedom.) For a real gas such as water vapour, these values are only approximate, but they give the correct order of magnitude. For example, the correct values are $c_P = 37.468$ joules per K (i.e., 4.5R) and $c_P - c_V = 9.443$ joules per K (i.e., 1.14R) for water vapour at 100 °C and 1 atmosphere pressure.

Entropy as an Exact Differential

Because the quantity $dS = d'Qmax/T$ is an exact differential, many other important relationships connecting the thermodynamic properties of substances can be derived. For example, with the substitutions $d'Q = T\,dS$ and $d'W = P\,dV$, the differential form ($dU = d'Q - d'W$) of the first law of thermodynamics becomes (for a single pure substance),

$$dU = T\,dS - P\,dV.$$

The advantage gained by the above formula is that dU is now expressed entirely in terms of state functions in place of the path-dependent quantities $d'Q$ and $d'W$. This change has the very important mathematical implication that the appropriate independent variables are S and V in place of T and V, respectively, for internal energy.

This replacement of T by S as the most appropriate independent variable for the internal energy of substances is the single most valuable insight provided by the combined first and second laws of thermodynamics. With U regarded as a function $U(S, V)$, its differential dU is,

$$\mathrm{d}U = \left(\frac{\partial U}{\partial S}\right)_V dS + \left(\frac{\partial U}{\partial V}\right)_S dV.$$

A comparison with the preceding equation shows immediately that the partial derivatives are,

$$\left(\frac{\partial U}{\partial S}\right)_V = T \text{ and } \left(\frac{\partial U}{\partial V}\right)_S = -P.$$

Furthermore, the cross partial derivatives,

$$\left(\frac{\partial^2 U}{\partial V \partial S}\right) = \left(\frac{\partial T}{\partial V}\right)_S \text{ and } \left(\frac{\partial^2 U}{\partial S \partial V}\right) = -\left(\frac{\partial P}{\partial S}\right)_V.$$

Must be equal because the order of differentiation in calculating the second derivatives of U does not matter. Equating the right-hand sides of the above pair of equations then yields,

$$\left(\frac{\partial T}{\partial V}\right)_S = -\left(\frac{\partial P}{\partial S}\right)_V.$$

This is one of four Maxwell relations (the others will follow shortly). They are all extremely useful in that the quantity on the right-hand side is virtually impossible to measure directly, while the quantity on the left-hand side is easily measured in the laboratory. For the present case one simply measures the adiabatic variation of temperature with volume in an insulated cylinder so that there is no heat flow (constant S).

The other three Maxwell relations follow by similarly considering the differential expressions for the thermodynamic potentials $F(T, V)$, $H(S, P)$, and $G(T, P)$, with independent variables as indicated. The results are,

$$\left(\frac{\partial P}{\partial T}\right)_V = \left(\frac{\partial S}{\partial V}\right)_T, \left(\frac{\partial V}{\partial S}\right)_P = \left(\frac{\partial T}{\partial P}\right)_S, \left(\frac{\partial V}{\partial T}\right)_P = -\left(\frac{\partial S}{\partial P}\right)_T$$

As an example of the use of these equations, equation $C_P = C_V + \left[P + \left(\frac{\partial U}{\partial V}\right)_T\right]\left(\frac{\partial V}{\partial T}\right)_P$. for $CP - CV$ contains the partial derivative,

$$\left(\frac{\partial U}{\partial V}\right)_T$$

which vanishes for an ideal gas and is difficult to evaluate directly from experimental data for real substances. The general properties of partial derivatives can first be used to write it in the form,

$$\left(\frac{\partial U}{\partial V}\right)_T = \left(\frac{\partial U}{\partial V}\right)_S + \left(\frac{\partial U}{\partial s}\right)_V \left(\frac{\partial S}{\partial V}\right)_T.$$

Combining this with equation $\left(\frac{\partial U}{\partial S}\right)_V = T$ and $\left(\frac{\partial U}{\partial V}\right)_S = -P$. for the partial derivatives together with the first of the Maxwell equations from equation $\left(\frac{\partial P}{\partial T}\right)_V = \left(\frac{\partial S}{\partial V}\right)_T$, $\left(\frac{\partial V}{\partial S}\right)_P = \left(\frac{\partial T}{\partial P}\right)_S$, $\left(\frac{\partial V}{\partial T}\right)_P = -\left(\frac{\partial S}{\partial P}\right)_T$ then yields the desired result,

$$\left(\frac{\partial U}{\partial V}\right)_T = -P + T\left(\frac{\partial P}{\partial T}\right)_V.$$

The quantity,

$$\left(\frac{\partial P}{\partial T}\right)_V$$

comes directly from differentiating the equation of state. For an ideal gas,

$$\left(\frac{\partial P}{\partial T}\right)_V = \frac{nR}{V} = \frac{P}{T},$$

so,

$$\left(\frac{\partial U}{\partial V}\right)_T$$

is zero as expected. The departure of, $\left(\frac{\partial U}{\partial V}\right)_T$ from zero reveals directly the effects of internal forces between the molecules of the substance and the work that must be done against them as the substance expands at constant temperature.

The Clausius-clapeyron Equation

Phase changes, such as the conversion of liquid water to steam, provide an important example of a system in which there is a large change in internal energy with volume at constant temperature. Suppose that the cylinder contains both water and steam in equilibrium with each other at pressure *P*, and the cylinder is held at constant temperature *T*, as shown in the figure. The pressure remains equal to the vapour pressure Pvap as the piston moves up, as long as both phases remain present. All that happens is that more water turns to steam, and the heat reservoir must supply the latent heat of vaporization, $\lambda = 40.65$ kilojoules per mole, in order to keep the temperature constant.

The results of the preceding section can be applied now to find the variation of the boiling point of water with pressure. Suppose that as the piston moves up, 1 mole of water turns to steam. The change in volume inside the cylinder is then $\Delta V = V_{gas} - V_{liquid}$, where $V_{gas} = 30.143$ litres is the volume of 1 mole of steam at 100 °C, and $V_{liquid} = 0.0188$ litre is the volume of 1 mole of water. By the first law of thermodynamics, the change in internal energy ΔU for the finite process at constant *P* and *T* is $\Delta U = \lambda - P\Delta V$.

The variation of *U* with volume at constant *T* for the complete system of water plus steam is thus,

$$\left(\frac{\partial U}{\partial V}\right)_T = \frac{\Delta U}{\Delta V} = \frac{\lambda}{\Delta V} - P.$$

A comparison with equation $\left(\frac{\partial U}{\partial V}\right)_T = -P + T\left(\frac{\partial P}{\partial T}\right)_V$ then yields the equation,

$$\left(\frac{\partial P}{\partial T}\right)_V = \frac{\lambda}{\Delta V}.$$

However, for the present problem, P_{vapour} is the vapour pressure vapour, which depends only on T and is independent of V. The partial derivative is then identical to the total derivative,

$$\frac{dP_{vapour}}{\partial T}$$

giving the Clausius - Clapeyron equation,

$$\left(\frac{dP}{\partial T}\right)_V = \frac{\lambda}{\Delta V}.$$

This equation is very useful because it gives the variation with temperature of the pressure at which water and steam are in equilibrium—i.e., the boiling temperature. An approximate but even more useful version of it can be obtained by neglecting V_{liquid} in comparison with V_{gas} and using,

$$V_{gas} = \frac{RT}{P_{vapour}}$$

from the ideal gas law. The resulting differential equation can be integrated to give,

$$\frac{1}{T} = \frac{1}{T_0} + \frac{R}{\lambda} \ln \frac{P_0}{P}$$

For example, at the top of Mount Everest, atmospheric pressure is about 30 percent of its value at sea level. Using the values $R = 8.3145$ joules per K and $\lambda = 40.65$ kilojoules per mole, the above equation gives T = 342 K (69 °C) for the boiling temperature of water, which is barely enough to make tea.

LAWS OF THERMODYNAMICS

Zeroth Law of Thermodynamics

The Zeroth Law of Thermodynamics states that systems in thermal equilibrium are at the same temperature.

There are a few ways to state the Zeroth Law of Thermodynamics, but the simplest is as follows: systems that are in thermal equilibrium exist at the same temperature.

Systems are in thermal equilibrium if they do not transfer heat, even though they are in a position to do so, based on other factors. For example, food that's been in the refrigerator overnight is in thermal equilibrium with the air in the refrigerator: heat no longer flows from one source (the food) to the other source (the air) or back.

What the Zeroth Law of Thermodynamics means is that temperature is something worth measuring, because it indicates whether heat will move between objects. This will be true regardless of how the objects interact. Even if two objects don't touch, heat may still flow between them, such as by radiation (as from a heat lamp). However, according to the Zeroth Law of Thermodynamics, if the systems are in thermal equilibrium, no heat flow will take place.

There are more formal ways to state the Zeroth Law of Thermodynamics, which is commonly stated in the following manner:

Let A, B, and C be three systems. If A and C are in thermal equilibrium, and A and B are in thermal equilibrium, then B and C are in thermal equilibrium.

This statement is represented symbolically in. Temperature is not mentioned explicitly, but it's implied that temperature exists. Temperature is the quantity that is always the same for all systems in thermal equilibrium with one another.

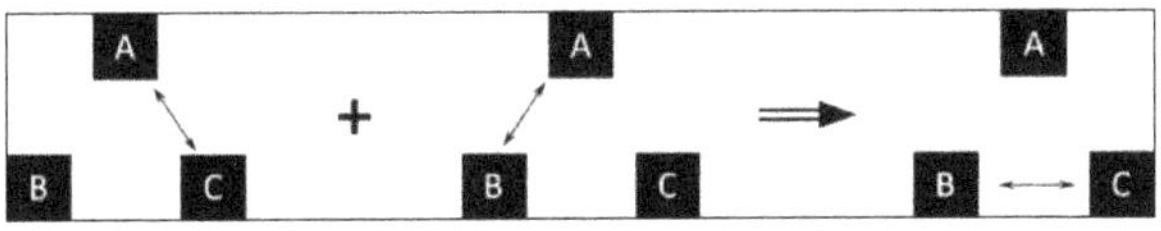

Zeroth Law of Thermodynamics: The double arrow represents thermal equilibrium between systems. If systems A and C are in equilibrium, and systems A and B are in equilibrium, then systems B and C are in equilibrium. The systems A, B, and C are at the same temperature.

First Law of Thermodynamics

Many power plants and engines operate by turning heat energy into work. The reason is that a heated gas can do work on mechanical turbines or pistons, causing them to move. The first law of thermodynamics applies the conservation of energy principle to systems where heat transfer and doing work are the methods of transferring energy into and out of the system. The first law of thermodynamics states that the change in internal energy of a system ΔU equals the net heat transfer into the system Q, plus the net work done on the system W. In equation form, the first law of thermodynamics is,

$$\Delta U = Q + W.$$

Here ΔU is the change in internal energy U of the system. Q is the net heat transferred into the system—that is, Q is the sum of all heat transfer into and out of the system. W is the net work done on the system.

So positive heat Q adds energy to the system and positive work WWW adds energy to the system. This is why the first law takes the form it does, $\Delta U = Q + W$ Q, plus, W. It simply says that you can add to the internal energy by heating a system, or doing work on the system.

Nothing quite exemplifies the first law of thermodynamics as well as a gas (like air or helium) trapped in a container with a tightly fitting movable piston (as seen below). We'll assume the piston can move up and down, compressing the gas or allowing the gas to expand (but no gas is allowed to escape the container).

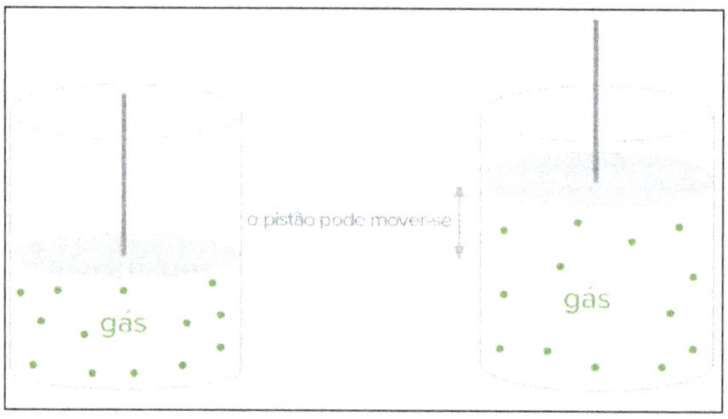

The gas molecules trapped in the container are the "system". Those gas molecules have kinetic energy.

The internal energy U of our system can be thought of as the sum of all the kinetic energies of the individual gas molecules. So, if the temperature T of the gas increases, the gas molecules speed up and the internal energy U of the gas increases (which means ΔU is positive). Similarly, if the temperature T of the gas decreases, the gas molecules slow down, and the internal energy UUU of the gas decreases (which means ΔU is negative).

It's really important to remember that internal energy U and temperature T will both increase when the speeds of the gas molecules increase, since they are really just two ways of measuring the same thing; how much energy is in a system. Since temperature and internal energy are proportional $T \propto U$, if the internal energy doubles the temperature doubles. Similarly, if the temperature does not change, the internal energy does not change.

One way we can increase the internal energy U (and therefore the temperature) of the gas is by transferring heat Q into the gas. We can do this by placing the container over a Bunsen burner or submerging it in boiling water. The high temperature environment would then conduct heat thermally through the walls of the container and into the gas, causing the gas molecules to move faster. If heat enters the gas, Q will be a positive number. Conversely, we can decrease the internal energy of the gas by transferring heat out of the gas. We could do this by placing the container in an ice bath. If heat exits the gas, Q will be a negative number. This sign convention for heat Q is represented in the image below.

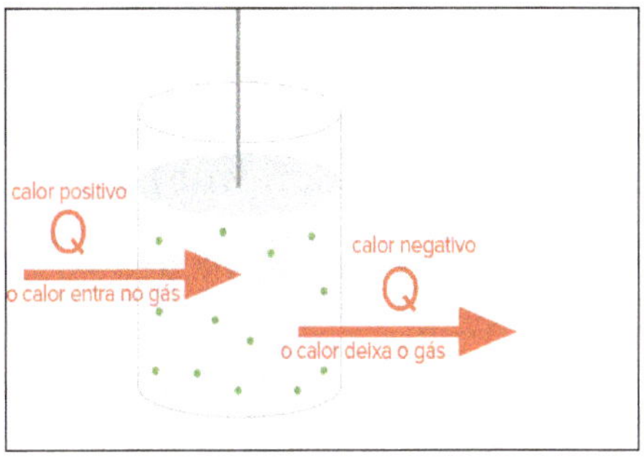

Since the piston can move, the piston can do work on the gas by moving downward and compressing the gas. The collision of the downward moving piston with the gas molecules causes the gas molecules to move faster, increasing the total internal energy. If the gas is compressed, the work done on the gas $W_{on\ gas}$ is a positive number. Conversely, if the gas expands and pushes the piston upward, work is done by the gas. The collision of the gas molecules with the receding piston causes the gas molecules to slow down, decreasing the internal energy of the gas. If the gas expands, the work done on the gas $W_{on\ gas}$ is a negative number. This sign convention for work W is represented in the image below.

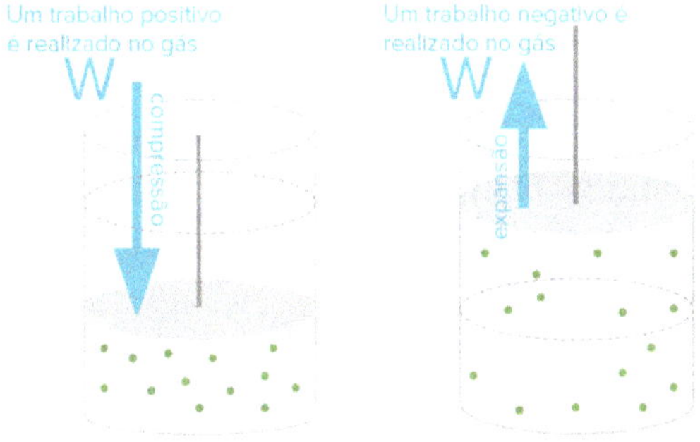

Below is a table that summarizes the signs conventions for all three quantities $(\Delta U, Q, W)$ left parenthesis, delta, U, comma, Q, comma, W, right parenthesis.

ΔU (change in internal energy)	Q (heat)	W (work done on gas)
is + plus if temperature T	is + plus if heat enters gas	is + plus if gas is compressed
is − minus if temperature T	is − minus if heat exits gas	is − minus if gas expands
is 0 if temperature T is constant	is 0 if no heat exchanged	is 0 if volume is constant

Is heat Q the same thing as temperature T ?

Absolutely not. This is one of the most common misconceptions when dealing with the first law of thermodynamics. The heat Q represents the heat energy that enters a gas (e.g. thermal conduction through the walls of the container). The temperature T on the other hand, is a number that's proportional to the total internal energy of the gas. So, Q is the energy a gas gains through thermal conduction, but T is proportional to the total amount of energy a gas has at a given moment. The heat that enters a gas might be zero $(Q = 0)$ left parenthesis, Q, equals, 0, right parenthesis if the container is thermally insulated, however, that does not mean that the temperature of the gas is zero (since the gas likely had some internal energy to start with).

To drive this point home, consider the fact that the temperature T of a gas can increase even if heat Q leaves the gas. This sounds counterintuitive, but since both work and heat can change the internal energy of a gas, they can both affect the temperature of a gas. For instance, if you place a piston in a sink of ice water, heat will conduct energy out the gas. However, if we compress the piston so that the work done on the gas is greater than the heat energy that leaves the gas, the total internal energy of the gas will increase.

Second Law of Thermodynamics

The Second Law of Thermodynamics states that the state of entropy of the entire universe, as an isolated system, will always increase over time. The second law also states that the changes in the entropy in the universe can never be negative.

Why is it that when you leave an ice cube at room temperature, it begins to melt? Why do we get older and never younger? And, why is it whenever rooms are cleaned, they become messy again in the future? Certain things happen in one direction and not the other, this is called the "arrow of time" and it encompasses every area of science. The thermodynamic arrow of time (entropy) is the measurement of disorder within a system. Denoted as ΔS, the change of entropy suggests that time itself is asymmetric with respect to order of an isolated system, meaning: a system will become more disordered, as time increases.

Major Players in Developing the Second Law

- Nicolas Léonard Sadi Carnot was a French physicist, who is considered to be the "father of thermodynamics," for he is responsible for the origins of the Second Law of Thermodynamics, as well as various other concepts. The current form of the second law uses entropy rather than caloric, which is what Sadi Carnot used to describe the law. Caloric relates to heat and Sadi Carnot came to realize that some caloric is always lost in the motion cycle. Thus, the thermodynamic reversibility concept was proven wrong, proving that irreversibility is the result of every system involving work.

- Rudolf Clausius was a German physicist, and he developed the Clausius statement, which says "Heat generally cannot flow spontaneously from a material at a lower temperature to a material at a higher temperature."

- William Thompson, also known as Lord Kelvin, formulated the Kelvin statement, which states "It is impossible to convert heat completely in a cyclic process." This means that there is no way for one to convert all the energy of a system into work, without losing energy.
- Constantin Carathéodory, a Greek mathematician, created his own statement of the second low arguing that "In the neighborhood of any initial state, there are states which cannot be approached arbitrarily close through adiabatic changes of state."

Probabilities

If a given state can be accomplished in more ways, then it is more probable than the state that can only be accomplished in a fewer/one way.

Assume a box filled with jigsaw pieces were jumbled in its box, the probability that a jigsaw piece will land randomly, away from where it fits perfectly, is very high. Almost every jigsaw piece will land somewhere away from its ideal position. The probability of a jigsaw piece landing correctly in its position, is very low, as it can only happened one way. Thus, the misplaced jigsaw pieces have a much higher multiplicity than the correctly placed jigsaw piece, and we can correctly assume the misplaced jigsaw pieces represent a higher entropy.

Derivation and Explanation

To understand why entropy increases and decreases, it is important to recognize that two changes in entropy have to considered at all times. The entropy change of the surroundings and the entropy change of the system itself. Given the entropy change of the universe is equivalent to the sums of the changes in entropy of the system and surroundings:

$$\Delta S_{univ} = \Delta S_{sys} + \Delta S_{sur}r = \frac{q_{sys}}{T} + \frac{q_{surr}}{T}$$

In an isothermal reversible expansion, the heat q absorbed by the system from the surroundings is,

$$q_{rev} = nRT\ ln\frac{V_2}{V_1}$$

Since the heat absorbed by the system is the amount lost by the surroundings, $q_{sys} = -q_{surr}$. Therefore, for a truly reversible process, the entropy change is,

$$\Delta S_{univ} = \frac{nRT\ ln\frac{V_2}{V_1}}{T} + \frac{-nRT\ ln\frac{V_2}{V_1}}{T} = 0$$

If the process is irreversible however, the entropy change is,

$$\Delta_{univ} = \frac{nRT\ ln—}{} >$$

If we put the two equations for ΔS_{univ} together for both types of processes, we are left with the second law of thermodynamics,

$$\Delta S_{univ} = \Delta S_{sys} + \Delta S_{surr} \geq 0$$

where ΔS_{univ} equals zero for a truly reversible process and is greater than zero for an irreversible process. In reality, however, truly reversible processes never happen (or will take an infinitely long time to happen), so it is safe to say all thermodynamic processes we encounter everyday are irreversible in the direction they occur.

Gibbs Free Energy

Given another equation:

$$\Delta S_{total} = \Delta S_{univ} = \Delta S_{surr} + \Delta S_{sys}$$

The formula for the entropy change in the surroundings is $\Delta S_{surr} = \Delta H_{sys} / T$. If this equation is replaced in the previous formula, and the equation is then multiplied by T and by -1 it results in the following formula.

$$-T\Delta S_{univ} = \Delta H_{sys} - T\Delta S_{sys}$$

If the left side of the equation is replaced G, which is know as Gibbs energy or free energy, the equation becomes,

$$\Delta G = \Delta H - T\Delta S$$

Now it is much simpler to conclude whether a system is spontaneous, non-spontaneous, or at equilibrium.

ΔH refers to the heat change for a reaction. A positive ΔH means that heat is taken from the environment (endothermic). A negative ΔH means that heat is emitted or given the environment (exothermic).

ΔG is a measure for the change of a system's free energy in which a reaction takes place at constant pressure (P) and temperature (T).

According to the equation, when the entropy decreases and enthalpy increases the free energy change, ΔG, is positive and not spontaneous, and it does not matter what the temperature of the system is. Temperature comes into play when the entropy and enthalpy both increase or both decrease. The reaction is not spontaneous when both entropy and enthalpy are positive and at low temperatures, and the reaction is spontaneous when both entropy and enthalpy are positive and at high temperatures. The reactions are spontaneous when the entropy and enthalpy are negative at low temperatures, and the reaction is not spontaneous when the entropy and enthalpy are negative at high temperatures. Because all spontaneous reactions increase entropy, one can determine if the entropy changes according to the spontaneous nature of the reaction.

Table: Matrix of Conditions Dictating Spontaneity.

Case	ΔH	ΔS	ΔG	Answer
High temperature	-	+	-	Spontaneous
Low temperature	-	+	-	Spontaneous
High temperature	-	-	+	Nonspontaneous
Low temperature	-	-	-	Spontaneous
High temperature	+	+	-	Spontaneous
Low temperature	+	+	+	Nonspontaneous
High temperature	+	-	+	Nonspontaneous
Low temperature	+	-	+	Nonspontaneous

Application of the Second Law

The second law occurs all around us all of the time, existing as the biggest, most powerful, general idea in all of science.

Explanation of Earth's Age

When scientists were trying to determine the age of the Earth during 1800s they failed to even come close to the value accepted today. They also were incapable of understanding how the earth transformed. Lord Kelvin, first hypothesized that the earth's surface was extremely hot, similar to the surface of the sun. He believed that the earth was cooling at a slow pace. Using this information, Kelvin used thermodynamics to come to the conclusion that the earth was at least twenty million years, for it would take about that long for the earth to cool to its current state. Twenty million years was not even close to the actual age of the Earth, but this is because scientists during Kelvin's time were not aware of radioactivity. Even though Kelvin was incorrect about the age of the planet, his use of the second law allowed him to predict a more accurate value than the other scientists at the time.

Evolution and the Second Law

Some critics claim that evolution violates the Second Law of Thermodynamics, because organization and complexity increases in evolution. However, this law is referring to isolated systems only, and the earth is not an isolated system or closed system. This is evident for constant energy increases on earth due to the heat coming from the sun. So, order may be becoming more organized, the universe as a whole becomes more disorganized for the sun releases energy and becomes disordered.

Third Law of Thermodynamics

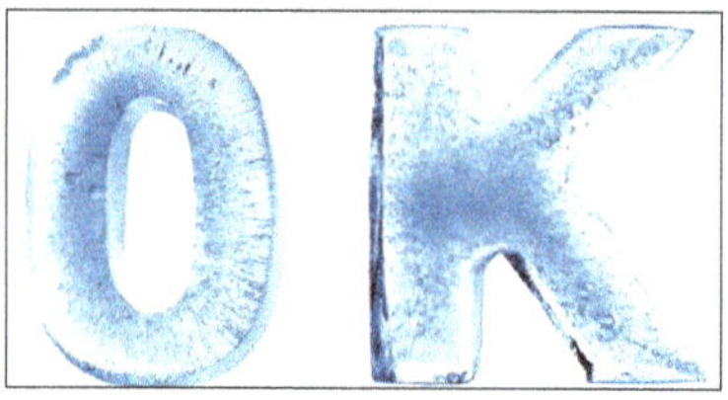

At 0 K, entropy stops. This is known as absolute zero, and in theory, this is not possible.

The Third Law of Thermodynamics is concerned with the limiting behavior of systems as the temperature approaches absolute zero. Most thermodynamics calculations use only entropy differences, so the zero point of the entropy scale is often not important. However, we discuss the Third Law for purposes of completeness because it describes the condition of zero entropy.

The Third Law states, "The entropy of a perfect crystal is zero when the temperature of the crystal is equal to absolute zero (0 K)." According to Purdue University, "The crystal must be perfect, or else there will be some inherent disorder. It also must be at 0 K; otherwise there will be thermal motion within the crystal, which leads to disorder."

"One version of the Third Law states that it would require an infinite number of steps to reach absolute zero, which means you will never get there. If you could get to absolute zero, it would violate the Second Law, because if you had a heat sink at absolute zero, then you could build a machine that was 100 percent efficient."

In theory it would be possible to grow a perfect crystal in which all of the lattice spaces are occupied by identical atoms. However, it is generally believed that it is impossible to achieve a temperature of absolute zero (although scientists have come quite close). Therefore, all matter contains at least some entropy owing to the presence of some heat energy.

The Third Law of Thermodynamics was first formulated by German chemist and physicist Walther Nernst. In his book, "A Survey of Thermodynamics", Martin Bailyn quotes Nernst's statement of the Third Law as, "It is impossible for any procedure to lead to the isotherm $T = 0$ in a finite number of steps." This essentially establishes a temperature absolute zero as being unattainable in somewhat the same way as the speed of light c. Theory states and experiments have shown that no matter how fast something is moving, it can always be made to go faster, but it can never reach the speed of light. Similarly, no matter how cold a system is, it can always be made colder, but it can never reach absolute zero.

In her book, "The Story of Physics", Anne Rooney wrote, "The third law of thermodynamics requires the concept of a minimum temperature below which no temperature can ever fall — known as absolute zero." She continued, "Robert Boyle first discussed the concept of a minimum possible temperature in 1665, in "New Experiments and Observations Touching Cold," in which he referred to the idea as primum frigidum."

Absolute zero is believed to have been first calculated with reasonable precision in 1779 by Johann Heinrich Lambert. He based this calculation on the linear relationship between the pressure and temperature of a gas. When a gas is heated in a confined space, its pressure increases. This is because the temperature of a gas is a measure of the average speed of the molecules in the gas. The hotter it gets, the faster the molecules move, and the greater the pressure they exert when they collide with the walls of the container. It was reasonable for Lambert to assume that if the temperature of the gas could be brought to absolute zero, the motion of the gas molecules could be brought to a complete stop so they could no longer exert any pressure on the walls of the chamber.

If one were to plot the temperature-pressure relationship of the gas on a graph with temperature on the x (horizontal) axis and pressure on the y (vertical) axis, the points form an upward-sloping straight line, indicating a linear relationship between temperature and pressure. It should be rather simple, then, to extend the line backward and read the temperature where the line crosses the xaxis, i.e., where $y = 0$, indicating zero pressure. Using this technique, Lambert calculated absolute zero to be minus 270 degrees Celsius (minus 454 Fahrenheit), which was remarkably close to the modern accepted value of minus 273.15 C (minus 459.67 F).

The Kelvin Temperature Scale

The person most associated with the concept of absolute zero is William Thomson, 1st Baron Kelvin. The temperature unit bearing his name, the kelvin (K), is the one most commonly used by scientists worldwide. Temperature increments in the Kelvin scale are the same size as in the Celsius scale, but because it starts at absolute zero, rather than the freezing point of water, it can be used directly in mathematical calculations, particularly in multiplication and division. For example, 100 K actually is twice as hot as 50 K. A sample of confined gas at 100 K also contains twice as much thermal energy, and it has twice the pressure as it would have at 50 K. Such calculations cannot be done using the Celsius or Fahrenheit scales, i.e., 100 C is not twice as hot as 50 C, nor is 100 F twice as hot as 50 F.

Implications of the Third Law

Because a temperature of absolute zero is physically unattainable, the Third Law may be restated to apply to the real world as: the entropy of a perfect crystal approaches zero as its temperature approaches absolute zero. We can extrapolate from experimental data that the entropy of a perfect crystal reaches zero at absolute zero, but we can never demonstrate this empirically.

According to David McKee, a professor of physics at Missouri Southern State University, "There's a field of ultra-low-temperature research, and every time you turn around there's a new record low. These days, nanokelvin (nK = 10^{-9} K) temperatures are reasonably easy to achieve, and everyone's now working on picokelvins (pK =, 10^{-12} K)." As of this writing, the record-low temperature was achieved 1999 by the YKI-group of the Low Temperature Laboratory at Aalto University in Finland. They cooled a piece of rhodium metal to 100 pK, or 100 trillionths of a degree Celsius above absolute zero besting the previous record of 280 pK set by them in 1993.

While a temperature of absolute zero does not exist in nature, and we cannot achieve it in the laboratory, the concept of absolute zero is critical for calculations involving temperature and entropy. Many measurements imply a relationship to some starting point. When we state a distance, we have to ask, distance from what? When we state a time, we have to ask, time since when? Defining the zero value on the temperature scale gives meaning to positive values on that scale. When a temperature is stated as 100 K, it means that the temperature is 100 K above absolute zero, which is twice as far above absolute zero as 50 K and half as far as 200 K.

On first reading, the Third Law seems rather simple and obvious. However, it serves and the final period at the end of a long and consequential story that fully describes the nature of heat and thermal energy.

References

- Heat, science: britannica.com, Retrieved 12 July, 2019
- Whatisht, heattransfer, simwiki, docs: simscale.com, Retrieved 11 January, 2019
- Thermodynamics, science: britannica.com, Retrieved 1 August, 2019
- The-zeroth-law-of-thermodynamics, chapter, boundless-physics: lumenlearning.com, Retrieved 5 March, 2019
- What-is-the-first-law-of-thermodynamics, laws-of-thermodynamics, thermodynamics, physics, science: khanacademy.org, Retrieved 8 June, 2019
- Second-law-of-thermodynamics, the-four-laws-of-thermodynamics, thermodynamics, supplemental-modules-(physical-and-theoretical-chemistry), physical-and-theoretical-chemistry-textbook-map: libretexts.org, Retrieved 3 May 2019
- 50942-third-law-thermodynamics: livescience.com, Retrieved 15 July, 2019

2

Radiation Heat Transfer

Electromagnetic radiation generated by the thermal motion of particles in matter is referred to as radiation heat transfer. Some of its basic concepts include black body radiation, Stefan–Boltzmann law, Planck's law, etc. This chapter discusses in detail all these concepts related to radiation heat transfer.

Radiation, energy transfer across a system boundary due to a ΔT, by the mechanism of photon emission or electromagnetic wave emission.

Because the mechanism of transmission is photon emission, unlike conduction and convection, there need be no intermediate matter to enable transmission.

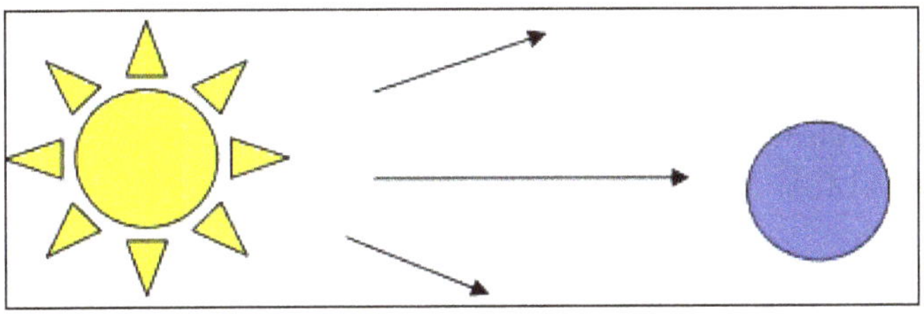

The significance of this is that radiation will be the only mechanism for heat transfer whenever a vacuum is present.

Electromagnetic Phenomena

We are well acquainted with a wide range of electromagnetic phenomena in modern life. These phenomena are sometimes thought of as wave phenomena and are, consequently, often described in terms of electromagnetic wave length, λ. Examples are given in terms of the wave distribution shown below:

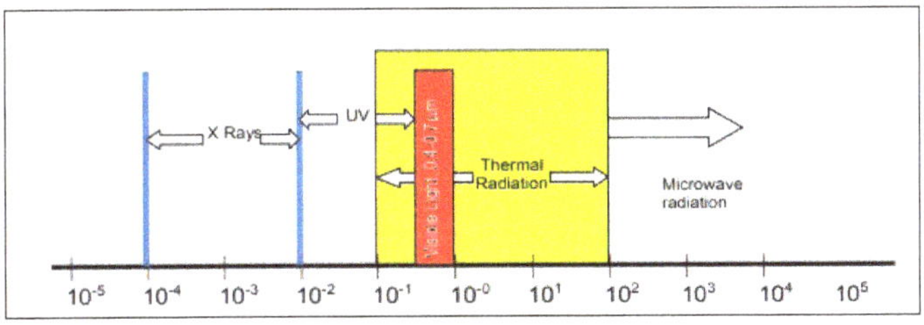

Wavelength, λ, μm.

One aspect of electromagnetic radiation is that the related topics are more closely associated with optics and electronics than with those normally found in mechanical engineering courses. Nevertheless, these are widely encountered topics and the student is familiar with them through everyday life experiences.

From a viewpoint of previously studied topics students, particularly those with a background in mechanical or chemical engineering, will find the subject of Radiation Heat Transfer a little unusual. The physics background differs fundamentally from that found in the areas of continuum mechanics. Much of the related material is found in courses more closely identified with quantum physics or electrical engineering, i.e. Fields and Waves. At this point, it is important for us to recognize that since the subject arises from a different area of physics, it will be important that we study these concepts with extra care.

Emission over Specific Wave Length Bands

Consider the problem of designing a tanning machine. As a part of the machine, we will need to design a very powerful incandescent light source. We may wish to know how much energy is being emitted over the ultraviolet band $\left(10^{-4}\ to\ 0.4\ \mu m\right)$, known to be particularly dangerous.

$$E_b\left(0.0001 \rightarrow 0.4\right) = \int_{0.001.\mu m}^{0.4\mu m} E_{b\lambda} \cdot d\lambda$$

With a computer available, evaluation of this integral is rather trivial. Alternatively, the text books provide a table of integrals. The format used is as follows:

$$\frac{E_b\left(0.0001 \rightarrow 0.4\right)}{E_b} = \frac{\int_{0.001.\mu m}^{0.4\mu m} E_{b\lambda} \cdot d\lambda}{\int_0^{\infty} E_{b\lambda} \cdot d\lambda} - \frac{\int_{0}^{0.0001.\mu m} E_{b\lambda} \cdot d\lambda}{\int_0^{\infty} E_{b\lambda} \cdot d\lambda} = F\left(0 \rightarrow 0.4\right) - F\left(0 \rightarrow 0.0001\right)$$

Referring to such tables, we see the last two functions listed in the second column. In the first column is a parameter, λ T. This is found by taking the product of the absolute temperature of the emitting surface, T, and the upper limit wave length, λ. In our example, suppose that the incandescent bulb is designed to operate at a temperature of 2000K. Reading from the table:

$\lambda,\ \mu m$	$T,\ K$	$\lambda.T,\ \mu m.K$	$F\left(0 \rightarrow \lambda\right)$
0.0001	2000	0.2	0
0.4	2000	600	0.000014
$F\left(0.4 \rightarrow 0.0001\mu m\right) = F\left(0 \rightarrow 0.4\mu m\right) - F\left(0 \rightarrow 0.0001\mu m\right)$			0.000014

This is the fraction of the total energy emitted which falls within the IR band. To find the absolute energy emitted multiply this value times the total energy emitted:

$$E_{bIR} = F\left(0.4 \rightarrow 0.0001\mu m\right) \cdot \sigma \cdot T^4 = 0.000014 \cdot 5.67 \cdot 10^{-8} \cdot 2000^4 = 12.7\ W/m^2$$

Solar Radiation

The magnitude of the energy leaving the Sun varies with time and is closely associated with such factors as solar flares and sunspots. Nevertheless, we often choose to work with an average value. The energy leaving the sun is emitted outward in all directions so that at any particular distance from the Sun we may imagine the energy being dispersed over an imaginary spherical area. Because this area increases with the distance squared, the solar flux also decreases with the distance squared. At the average distance between Earth and Sun this heat flux is 1353 W/m2, so that the average heat flux on any object in Earth orbit is found as:

$$G_{s,o} = S_c \cdot f \cdot \cos\theta.$$

Where,

S_c = Solar Constant, 1353 W/m2,

f = correction factor for eccentricity in Earth Orbit, $(0.97<f<1.03)$,

θ = Angle of surface from normal to Sun.

Because of reflection and absorption in the Earth's atmosphere, this number is significantly reduced at ground level. Nevertheless, this value gives us some opportunity to estimate the potential for using solar energy, such as in photovoltaic cells.

Angles and Arc Length

We are well accustomed to thinking of an angle as a two dimensional object. It may be used to find an arc length:

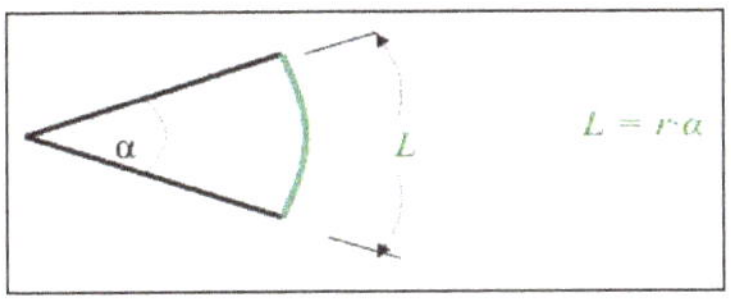

Solid Angle

We generalize the idea of an angle and an arc length to three dimensions and define a solid angle, Ω, which like the standard angle has no dimensions. The solid angle, when multiplied by the radius squared will have dimensions of length squared, or area, and will have the magnitude of the encompassed area.

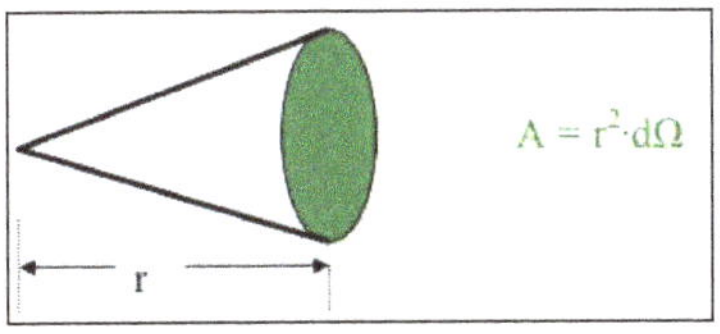

Projected Area

The area, dA_1, as seen from the prospective of a viewer, situated at an angle θ from the normal to the surface, will appear somewhat smaller, as $\cos\theta{\cdot}dA_1$. This smaller area is termed the projected area.

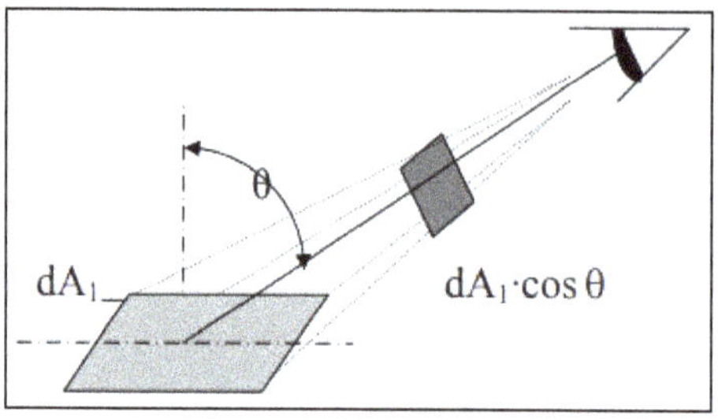

$$A_{projected} = cos\ \theta{\cdot}A_{normal}$$

Intensity

The ideal intensity, I_b, may now be defined as the energy emitted from an ideal body, per unit projected area, per unit time, per unit solid angle.

$$I = \frac{dq}{\cos\theta\cdot dA_1\cdot d\Omega}$$

Spherical Geometry

Since any surface will emit radiation outward in all directions above the surface, the spherical coordinate system provides a convenient tool for analysis. The three basic coordinates shown are R, φ, and θ, representing the radial, azimuthal and zenith directions.

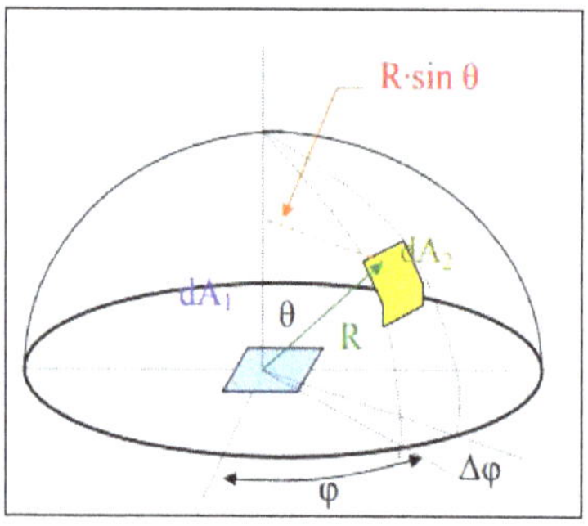

$$dA_2 = \left[(R\cdot\sin\theta)\cdot d\varphi\right].\left[R\cdot d\theta\right]$$

or, more simply as:

$$dA_2 = R^2\cdot\sin\theta{\cdot}\mathrm{d}\varphi{\cdot}\mathrm{d}\theta]$$

Recalling the definition of the solid angle,

$$dA = R^2 \cdot d\Omega$$

we find that:

$$d\Omega = R^2 \cdot \sin\theta \cdot d\theta \cdot d\varphi$$

Real Surfaces

Thus far we have spoken of ideal surfaces, i.e. those that emit energy according to the Stefan-Boltzman law:

$$E_b = \sigma \cdot T_{abs}{}^4$$

Real surfaces have emissive powers, E, which are somewhat less than that obtained theoretically by Boltzman. To account for this reduction, we introduce the emissivity, ε.

$$\varepsilon \equiv \frac{E}{E_b}$$

so that the emissive power from any real surface is given by:

$$E = \varepsilon \cdot \sigma \cdot T_{abs}{}^4$$

Receiving Properties

Targets receive radiation in one of three ways; they absorption, reflection or transmission. To account for these characteristics, we introduce three additional properties:

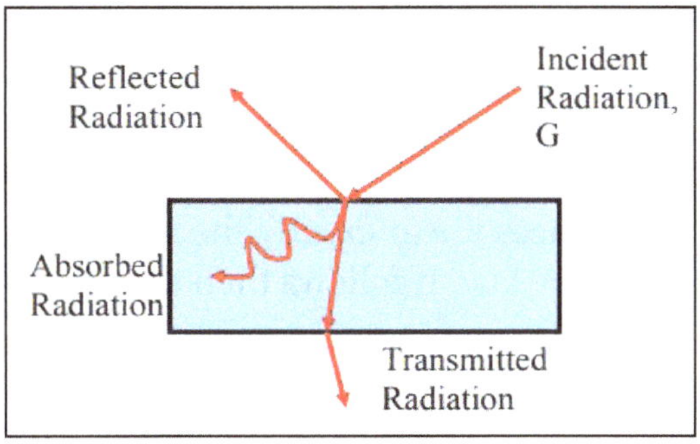

- Absorptivity, α, the fraction of incident radiation absorbed. Transmitted Radiation Absorbed Radiation Incident Radiation, G Reflected Radiation.
- Reflectivity, ρ, the fraction of incident radiation reflected.
- Transmissivity, τ, the fraction of incident radiation transmitted.

We see, from Conservation of Energy, that:

$$\alpha + \rho + \tau = 1$$

In this course, we will deal with only opaque surfaces, $\tau = 0$, so that:

$$\alpha + \rho = 1 \text{ Opaque Surfaces}$$

Relationship between Absorptivity and Emissivity

Consider two flat, infinite planes, surface A and surface B, both emitting radiation toward one another. Surface B is assumed to be an ideal emitter, i.e. $\varepsilon_B = 1.0$. Surface A will emit radiation according to the Stefan-Boltzman law as:

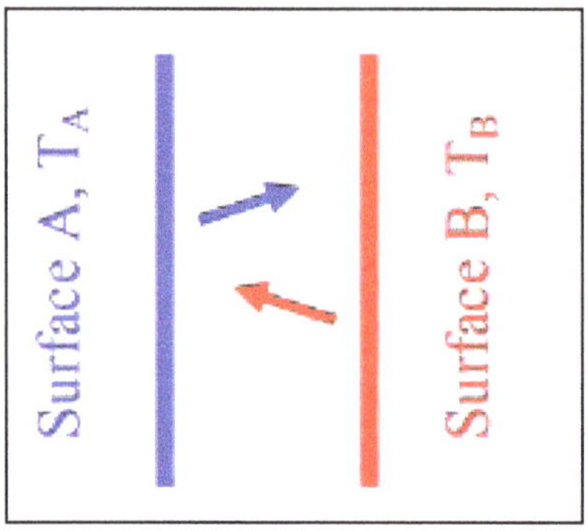

$$E_A = \varepsilon_A \cdot \sigma \cdot T_A^4$$

and will receive radiation as:

$$G_A = \alpha_A \cdot \sigma \cdot T_B^4$$

The net heat flow from surface A will be:

$$q'' = \varepsilon_A \cdot \sigma T_A^4 - \alpha_A \cdot \sigma T_B^4$$

Now suppose that the two surfaces are at exactly the same temperature. The heat flow must be zero according to the 2nd law. If follows then that:

$$\alpha A = \varepsilon A$$

Because of this close relation between emissivity, ε, and absorptivity, α, only one property is normally measured and this value may be used alternatively for either property.

Let's not lose sight of the fact that, as thermodynamic properties of the material, á and å may depend on temperature. In general, this will be the case as radiative properties will depend on wavelength, λ. The wave length of radiation will, in turn, depend on the temperature of the source of radiation.

The emissivity, ε, of surface A will depend on the material of which surface A is composed, i.e. aluminum, brass, steel, etc. and on the temperature of surface A.

The absorptivity, α, of surface A will depend on the material of which surface A is composed, i.e. aluminum, brass, steel, etc. and on the temperature of surface B.

In the design of solar collectors, engineers have long sought a material which would absorb all solar radiation, $\left(\alpha = 1, T_{sun} \sim 5600K\right)$ but would not re-radiate energy as it came to temperature $\left(\varepsilon=1, T_{collector} \sim 5600K\right)$. NASA developed an anodized chrome, commonly called "black chrome" as a result of this research.

Within the visual band of radiation, any material, which absorbs all visible light, appears as black. Extending this concept to the much broader thermal band, we speak of surfaces with α=1 as also being "black" or "thermally black". It follows that for such a surface, ε=1 and the surface will behave as an ideal emitter. The terms ideal surface and black surface are used interchangeably.

Lambert's Cosine Law

A surface is said to obey Lambert's cosine law if the intensity, I, is uniform in all directions. This is an idealization of real surfaces as seen by the emissivity at different zenith angles:

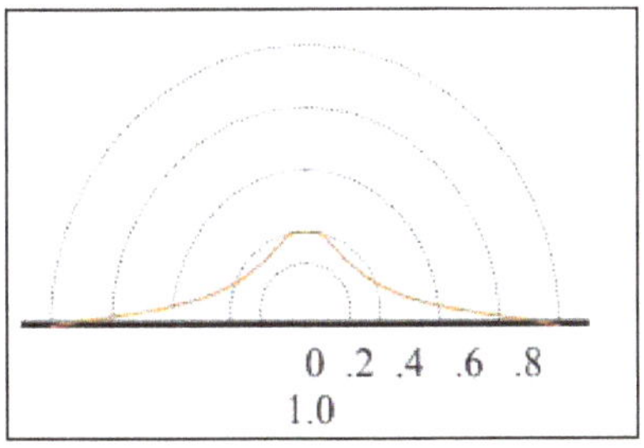

Dependence of Emissivity on Zenith Angle, Typical Metal.

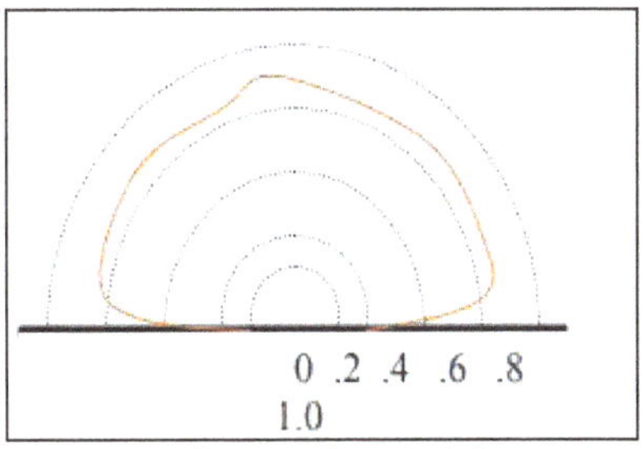

Dependence of Emissivity on Zenith Angle, Typical Non-Metal.

The sketches shown are intended to show is that metals typically have a very low emissivity, ε, which also remain nearly constant, expect at very high zenith angles, θ. Conversely, non-metals will have a relatively high emissivity, ε, except at very high zenith angles. Treating the emissivity as a constant over all angles is generally a good approximation and greatly simplifies engineering calculations.

Relationship Between Emissive Power and Intensity

By definition of the two terms, emissive power for an ideal surface, E_b, and intensity for an ideal surface, I_b.

$$E_b = \int_{hemisphere} I_b \cdot \cos\theta \cdot d\Omega$$

Replacing the solid angle by its equivalent in spherical angles:

$$E_b = \int_0^{2\cdot\pi}\int_0^{\pi/2} I_b \cdot \cos\theta \cdot \sin\theta \cdot d\theta \cdot d\varphi$$

Integrate once, holding I_b constant:

$$E_b = 2\cdot\pi\cdot I_b \cdot \int_0^{\pi/2} \cos\theta \cdot \sin\theta \cdot d\theta$$

Integrate a second time.

$$E_b = 2\cdot\pi\cdot I_b \cdot \left.\frac{\sin^2\theta}{2}\right|_0^{\pi/2} = \pi\cdot I_b$$

$$Eb = \pi\cdot I_b$$

Radiation Exchange

The intensity, I, herein is used to describe radiation within a particular solid angle.

$$I = \frac{dq}{\cos\theta \cdot dA_1 \cdot d\Omega}$$

This will now be used to determine the fraction of radiation leaving a given surface and striking a second surface.

Rearranging the above equation to express the heat radiated: $dq = I \cdot \cos\theta \cdot dA_1 \cdot d\Omega$

Next we will project the receiving surface onto the hemisphere surrounding the source. First find the projected area of surface dA_2, $dA_2 \cdot \cos\theta_2$. ($\theta 2$ is the angle between the normal to surface 2 and the position vector, R.) Then find the solid angle, Ω, which encompasses this area.

Substituting into the heat flow equation above: dA_2

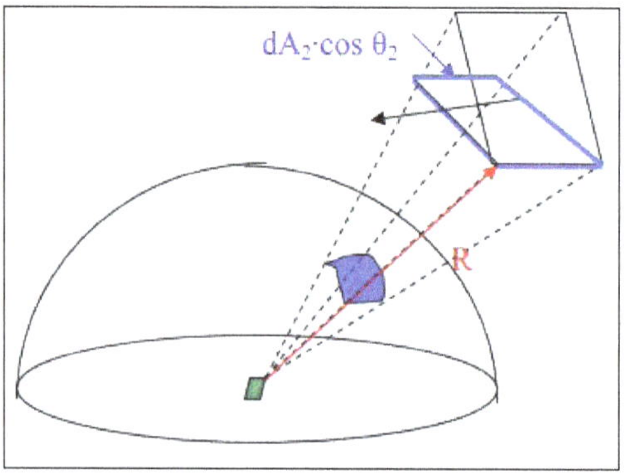

Substituting into the heat flow equation above:

$$dq = \frac{I \cdot \cos\theta_1 \cdot dA_1 \cdot \cos\theta_2 dA_2}{R^2}$$

To obtain the entire heat transferred from a finite area, dA_1, to a finite area, dA_2, we integrate over both surfaces:

$$q_{1\to 2} = \int_{A_2}\int_{A_1} \frac{I \cdot \cos\theta_1 \cdot dA_1 \cdot \cos\theta_2 dA_2}{R^2}$$

To express the total energy emitted from surface 1, we recall the relation between emissive power, E, and intensity, I.

$$q_{emitted} = E_1 \cdot A_1 = \pi \cdot I_1 \cdot A_1$$

View Factors-integral Method

Define the view factor, F_{1-2}, as the fraction of energy emitted from surface 1, which directly strikes surface 2.

$$F_{1\to 2} = \frac{q_{1\to 2}}{q_{emitted}} = \frac{\int_{A_2}\int_{A_1} \frac{I \cdot \cos\theta_1 \cdot dA_1 \cdot \cos\theta_2 dA_2}{R^2}}{\pi \cdot I_1 \cdot A_1}$$

after algebraic simplification this becomes:

$$F_{1\to 2} = \frac{1}{A_1} \cdot \int_{A_2}\int_{A_1} \frac{\cos\theta_1 \cdot \cos\theta_2 \cdot dA_1 \cdot dA_2}{\pi \cdot R^2}$$

Example Consider a diffuse circular disk of diameter D and area A_j and a plane diffuse surface of area $A_i << A_j$. The surfaces are parallel, and Ai is located at a distance L from the center of A_j. Obtain an expression for the view factor F_{ij}.

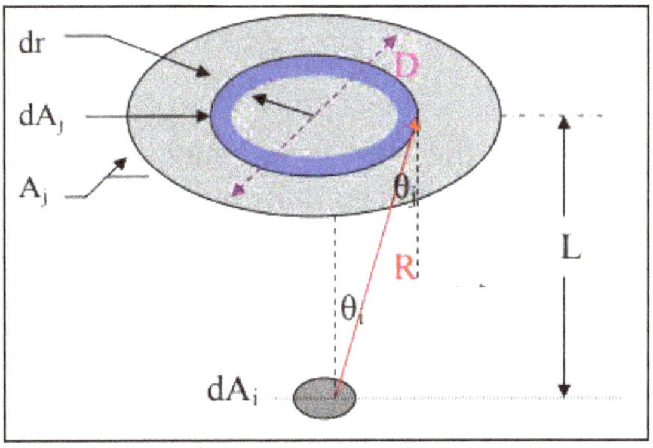

The view factor may be obtained from:

$$F_{1\to 2} = \frac{1}{A_1} \cdot \int_{A_2}\int_{A_1} \frac{\cos\theta_1 \cdot \cos\theta_2 \cdot dA_1 \cdot dA_2}{\pi \cdot R^2}$$

Since dA_i is a differential area,

$$F_{1\to 2} = \int_{A_1} \frac{\cos\theta_1 \cdot \cos\theta_2 \cdot dA_1 \cdot}{\pi \cdot R^2}$$

Substituting for the cosines and the differential area:

$$F_{1\to 2} = \int_{A_1} \frac{\left(L/R\right)^2 \cdot 2\pi \cdot r \cdot dr}{\pi \cdot R^2}$$

After simplifying:

$$F_{1\to 2} = \int_{A_1} \frac{L^2 \cdot 2\pi \cdot r \cdot dr}{R^4}$$

Let $\rho^2 = L^2 + r^2 = R^2$. Then $2 \cdot \rho \cdot d\rho = 2 \cdot r \cdot dr$.

$$F_{1\to 2} = \int_{A_1} \frac{L^2 \cdot 2 \cdot \rho \cdot d\rho}{\rho^4}$$

After integrating,

$$F_{1\to 2} = -2 \cdot L^2 \cdot \frac{\rho^{-2}}{2} = -L^2 \cdot \left[\frac{1}{L^2 + \rho^2}\right]_0^{D/2}$$

Substituting the upper & lower limits,

$$F_{1\to 2} = -L^2 \cdot \left[\frac{4}{4 \cdot L^2 + D^2} - \frac{1}{L^2}\right]_0^{D/2} = \frac{D^2}{4 \cdot L^2 + D^2}$$

This is but one example of how the view factor may be evaluated using the integral method. The approach used here is conceptually quite straight forward; evaluating the integrals and algebraically simplifying the resulting equations can be quite lengthy.

Enclosures

In order that we might apply conservation of energy to the radiation process, we must account for all energy leaving a surface. We imagine that the surrounding surfaces act as an enclosure about the heat source which receive all emitted energy. Should there be an opening in this enclosure through which energy might be lost, we place an imaginary surface across this opening to intercept this portion of the emitted energy. For an N surfaced enclosure, we can then see that:

$$\sum_{j=1}^{N} F_{i,j} = 1$$

This relationship is known as the "Conservation Rule".

Example: Consider the previous problem of a small disk radiating to a larger disk placed directly above at a distance L.

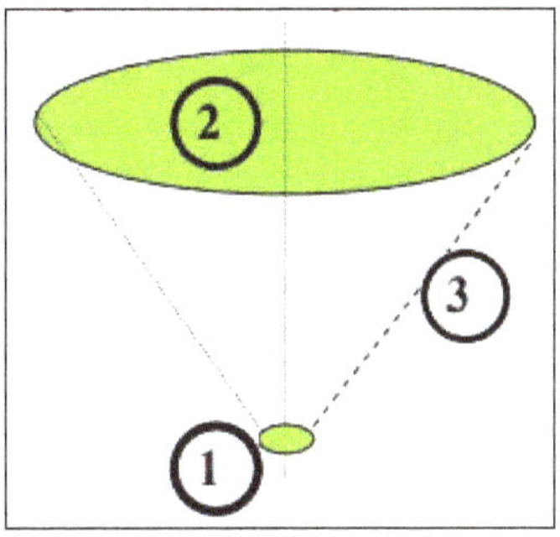

The view factor was shown to be 2 given by the relationship:

$$F_{1\to 2} = \frac{D^2}{4 \cdot L^2 + D^2}$$

Here, in order to provide an enclosure, we will define an imaginary surface 3, a truncated cone intersecting circles 1 and 2.

From our conservation rule we have:

$$\sum_{j=1}^{N} F_{i,j} = F_{1,1} + F_{1,2} + F_{1,3}$$

Since surface 1 is not convex $F_{1,1} = 0$. Then:

$$F_{1\to 3} = 1 - \frac{D^2}{4 \cdot L^2 + D^2}$$

Reciprocity

We may write the view factor from surface i to surface j as:

$$A_i \cdot F_{i\to j} = \int_{A_j}\int_{A_i} \frac{\cos\theta_i \cdot \cos\theta_j \cdot dA_i \cdot dA_j}{\pi \cdot R^2}$$

Similarly, between surfaces j and i:

$$A_j \cdot F_{j\to i} = \int_{A_j}\int_{A_i} \frac{\cos\theta_j \cdot \cos\theta_i \cdot dA_j \cdot dA_i}{\pi \cdot R^2}$$

Comparing the integrals we see that they are identical so that:

$$A_i \cdot F_{i\to j} = A_j \cdot F_{j\to i}$$

This relationship is known as "Reciprocity".

Example: Consider two concentric spheres shown to the right. All radiation leaving the outside of surface 1 will strike surface 2. Part of the radiant energy leaving the inside surface of object 2 will strike surface 1, part will return to surface 2. To find the fraction of energy leaving surface 2 which strikes surface 1, we apply reciprocity:

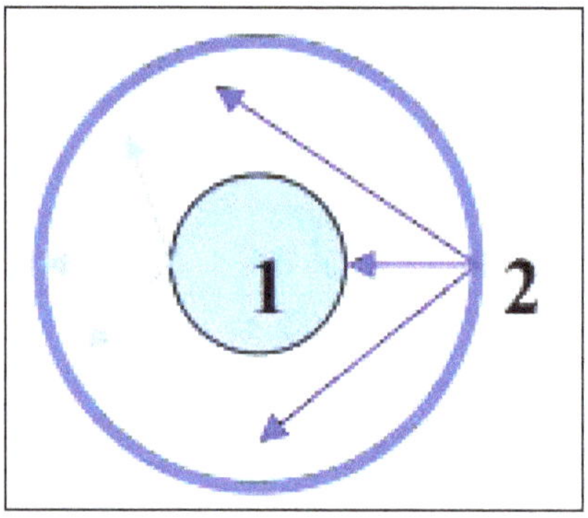

$$A_2 \cdot F_{2,1} = A_1 \cdot F_{1,2} \Rightarrow F_{2,1} = \frac{A_1}{A_2} \cdot F_{1,2} = \frac{A_1}{A_2} = \frac{D_1}{D_2}$$

Associative Rule

Consider the set of surfaces shown to the right: Clearly, from conservation of energy, the fraction of energy leaving surface i and striking the combined surface j+k will equal the fraction of energy emitted from i and striking j plus the fraction leaving surface i and striking k.

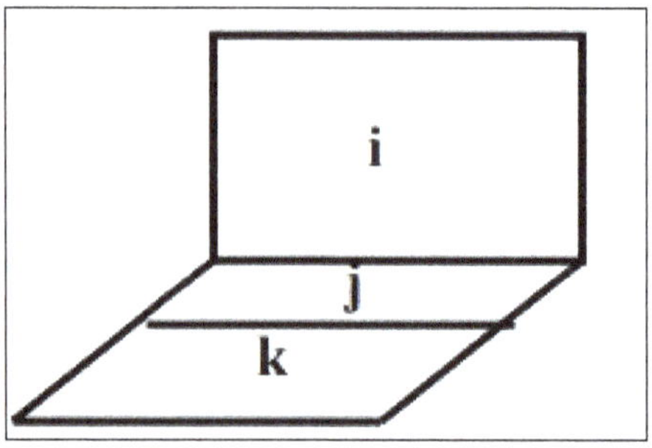

$$\sum_{j=1}^{N} F_{i,j} \quad F_{i\Rightarrow(j+k)} = F_{i\Rightarrow j} + F_{i\Rightarrow k}$$

This relationship is known as the "Associative Rule".

Radiosity

We have developed the concept of intensity, I, which let to the concept of the view factor. We have discussed various methods of finding view factors. There remains one additional concept to introduce before we can consider the solution of radiation problems.

Radiosity, J, is defined as the total energy leaving a surface per unit area and per unit time. This may initially sound much like the definition of emissive power, but the sketch below will help to clarify the concept.

$$J \equiv \varepsilon \cdot E_b + \rho \cdot G$$

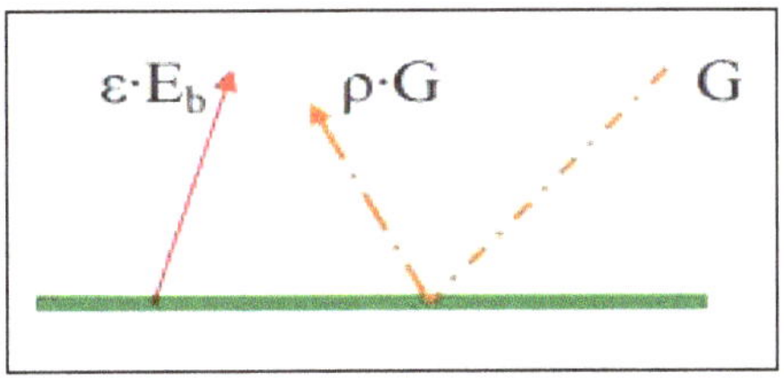

Net Exchange between Surfaces

Consider the two surfaces shown. Radiation will travel from surface i to surface j and will also travel from j to i.

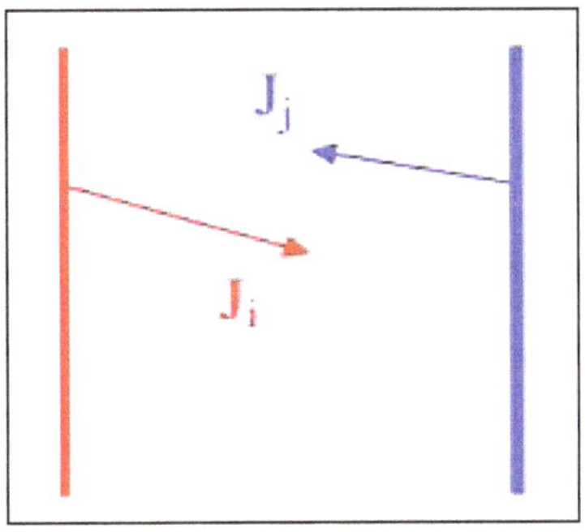

$$q_{i\rightarrow j} = J_i \cdot A_i \cdot F_{i\rightarrow j}$$

Likewise,

$$q_{j\rightarrow i} = J_j \cdot A_j \cdot F_{j\rightarrow j}$$

The net heat transfer is then:

$$q_{j\to i(net)} = J_i \cdot A_i \cdot J_{i\to j} - J_j \cdot A_j \cdot F_{j\to j}$$

From reciprocity we note that $F_{1\to 2} \cdot A_1 = F_{2\to 1} \cdot A_2$ so that,

$$q_{j\to i\,(net)} = J_i \cdot A_i \cdot F_{i\to j} - J_j \cdot A_i \cdot F_{i\to j} = A_i \cdot F_{i\to j} \cdot (Ji - Jj)$$

Net Energy Leaving a Surface

The net energy leaving a surface will be the difference between the energy leaving a surface and the energy received by a surface:

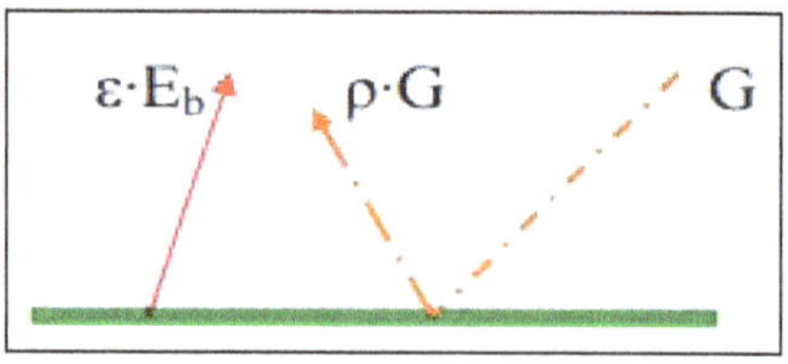

$$q_{1\to} = [\varepsilon \cdot E_b - \alpha \cdot G] \cdot A_1$$

Combine this relationship with the definition of Radiosity to eliminate G.

$$J \equiv \varepsilon \cdot E_b + \rho \cdot G \rightarrow G = [J - \varepsilon \cdot E_b]/\rho$$

$$q_{1\to} = \{\varepsilon \cdot E_b - \alpha \cdot [J - \varepsilon \cdot E_b]/\rho\} \cdot A_1$$

Assume opaque surfaces so that $\alpha + \rho = 1 \rightarrow \rho = 1 - \alpha$, and substitute for,

$$\rho.\ q_{1\to} = \{\varepsilon \cdot E_b - \alpha \cdot [J - \varepsilon \cdot E_b]/(1-\alpha)\} \cdot A_1$$

Put the equation over a common denominator:

$$q_{1\to} = \left[\frac{(1-\alpha) \cdot \varepsilon \cdot E_b - \alpha \cdot J + \alpha \cdot \varepsilon \cdot E_b}{1-\alpha}\right] \cdot A_1 = \left[\frac{\varepsilon \cdot E_b - \alpha \cdot J}{1-\alpha}\right] \cdot A_1$$

If we assume that $\alpha = \varepsilon$ then the equation reduces to:

$$q_{1\to} = \left[\frac{\varepsilon \cdot E_b - \alpha \cdot J}{1-\varepsilon}\right] \cdot A_1 = \left[\frac{\varepsilon \cdot A_1}{1-\varepsilon}\right] \cdot (E_b - J)$$

Electrical Analogy for Radiation We may develop an electrical analogy for radiation, similar to that produced for conduction. The two analogies should not be mixed: they have different dimensions on the potential differences, resistance and current flows.

	Equivalent Current	Equivalent Resistance	Potential Difference
Ohms Law	I	R	ΔV
Net Energy Leaving Surface	$q_{1\rightarrow}$	$\left[\dfrac{1-\varepsilon}{\varepsilon . A}\right]$	Eb - J
Net Exchange Between Surfaces	$q_{i\rightarrow j}$	$\dfrac{1}{A_1 \cdot F_{1\rightarrow 2}}$	J1 – J2

Alternate Procedure for Developing Networks

- Count the number of surfaces. (A surface must be at a "uniform" temperature and have uniform properties, i.e. $\varepsilon, \alpha, \rho$.)
- Draw a radiosity node for each surface.
- Connect the Radiosity nodes using view factor resistances $1/A_i \cdot F_{i\rightarrow j}$.
- Connect each Radiosity node to a grounded battery, through a surface resistance, $\left[1-\varepsilon/_{\varepsilon \cdot A}\right]$.

This procedure should lead to exactly the same circuit as we obtain previously.

Simplifications to the Electrical Network

- Insulated surfaces. In steady state heat transfer, a surface cannot receive net energy if it is insulated. Because the energy cannot be stored by a surface in steady state, all energy must be re-radiated back into the enclosure. Insulated surfaces are often termed as reradiating surfaces.

  Electrically cannot flow through a battery if it is not grounded.

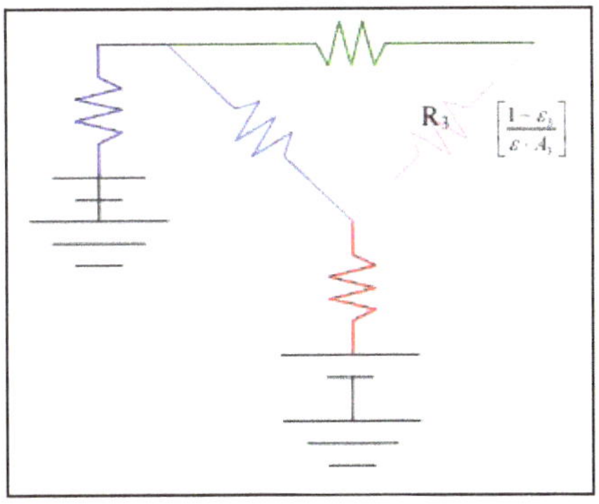

Surface 3 is not grounded so that the battery and surface resistance serve no purpose and are removed from the drawing.

- Black surfaces: A black, or ideal surface, will have no surface resistance:

$$\left[\frac{1-\varepsilon}{\varepsilon.A}\right]=\left[\frac{1-1}{1.A}\right]=0$$

In this case the nodal Radiosity and emissive power will be equal.

This result gives some insight into the physical meaning of a black surface. Ideal surfaces radiate at the maximum possible level. Nonblack surfaces will have a reduced potential, somewhat like a battery with a corroded terminal. They therefore have a reduced potential to cause heat/current flow.

- Large surfaces: Surfaces having a large surface area will behave as black surfaces, irrespective of the actual surface properties:

$$\left[\frac{1-\varepsilon}{\varepsilon.A}\right]=\left[\frac{1-\varepsilon}{\varepsilon.\infty}\right]=0$$

Physically, this corresponds to the characteristic of large surfaces that as they reflect energy, there is very little chance that energy will strike the smaller surfaces; most of the energy is reflected back to another part of the same large surface. After several partial absorptions most of the energy received is absorbed.

Solution of Analogous Electrical Circuits

- Large Enclosures: Consider the case of an object, 1, placed inside a large enclosure, 2. The system will consist of two objects, so we proceed to construct a circuit with two radiosity nodes.

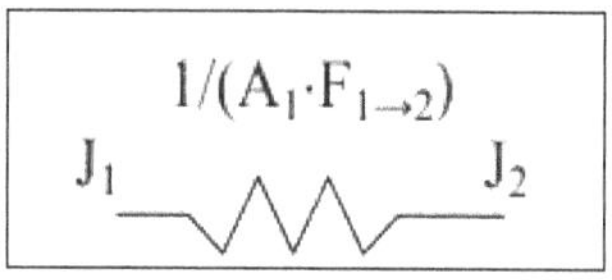

Now we ground both Radiosity nodes through a surface resistance.

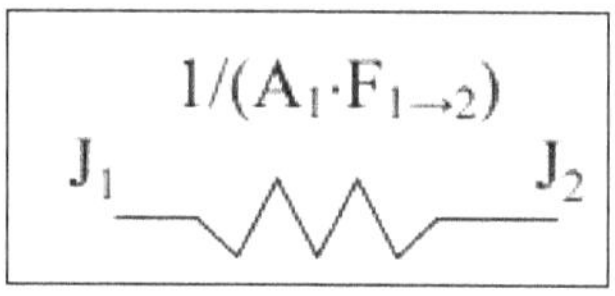

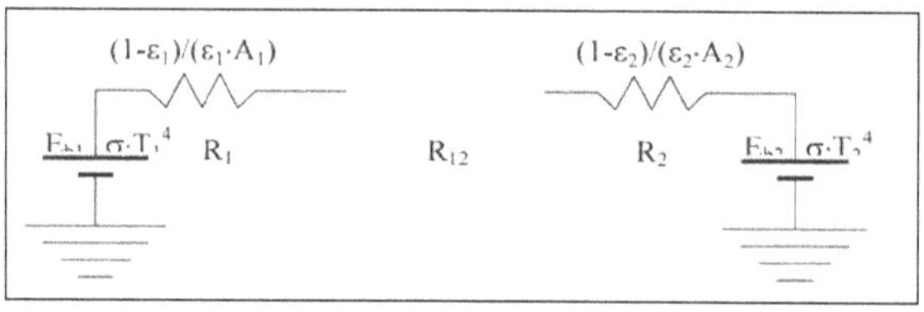

Since A_2 is large, $R_2 = 0$. The view factor, $F_{1\to2} = 1$.

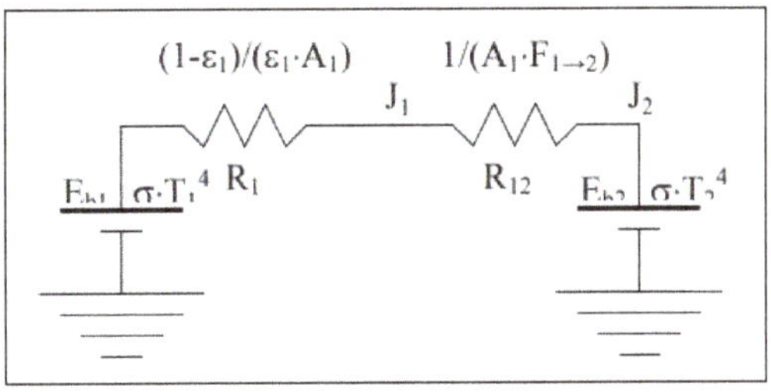

Sum the series resistances:

$$R_{Series} = (1-\varepsilon_1)/(\varepsilon_1 \cdot A_1) + 1/A_1 = 1/(\varepsilon_1 \cdot A_1)$$

Ohm's law:

$$i = \Delta V/R$$

or by analogy:

$$q = \Delta E_b/R_{Series} = \varepsilon_1 \cdot A_1 \cdot \sigma \cdot \left(T_1^4 - T_2^4\right)$$

You may recall this result from Thermo I, where it was introduced to solve this type of radiation problem.

- Networks with Multiple Potentials Systems with 3 or more grounded potentials will require a slightly different solution, but one which students have previously encountered in the Circuits course.

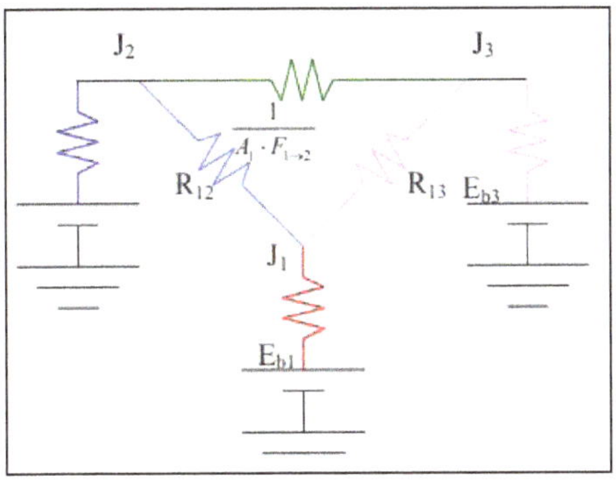

The procedure will be to apply Kirchoff's law to each of the Radiosity junctions.

$$\sum_{i=1}^{3} q_i = 0$$

In this example there are three junctions, so we will obtain three equations. This will allow us to solve for three unknowns.

Radiation problems will generally be presented on one of two ways:

- The surface net heat flow is given and the surface temperature is to be found.
- The surface temperature is given and the net heat flow is to be found.

Returning for a moment to the coal grate furnace, let us assume that we know (a) the total heat being produced by the coal bed, (b) the temperatures of the water walls and (c) the temperature of the super heater sections.

Apply Kirchoff's law about node 1, for the coal bed:

$$q_1 + q_{2\to1} + q_{3\to1} = q_1 + \frac{J_2 - J_1}{R_{12}} + \frac{J_3 - J_1}{R_{13}} = 0$$

Similarly, for node 2:

$$q_2 + q_{1\to2} + q_{3\to2} = \frac{E_{b2} - J_2}{R_2} + \frac{J_2 - J_1}{R_{12}} + \frac{J_3 - J_1}{R_{13}} = 0$$

And for node 3:

$$q_3 + q_{1\to3} + q_{2\to3} = \frac{E_{b3} - J_3}{R_3} + \frac{J_1 - J_3}{R_{13}} + \frac{J_2 - J_3}{R_{23}} = 0$$

The three equations must be solved simultaneously. Since they are each linear in J, matrix methods may be used:

$$\begin{bmatrix} -\frac{1}{R_{12}} - \frac{1}{R_{13}} & \frac{1}{R_{12}} & \frac{1}{R_{13}} \\ \frac{1}{R_{12}} & -\frac{1}{R_2} - \frac{1}{R_{12}} - \frac{1}{R_{13}} & \frac{1}{R_{23}} \\ \frac{1}{R_{13}} & \frac{1}{R_{23}} & -\frac{1}{R_2} - \frac{1}{R_{12}} - \frac{1}{R_{13}} \end{bmatrix} \cdot \begin{bmatrix} J_1 \\ J_2 \\ J_3 \end{bmatrix} = \begin{bmatrix} -q_1 \\ -\frac{E_{b2}}{R_2} \\ -\frac{E_{b3}}{R_3} \end{bmatrix}$$

The matrix may be solved for the individual Radiosity. Once these are known, we return to the electrical analogy to find the temperature of surface 1, and the heat flows to surfaces 2 and 3.

Surface 1: Find the coal bed temperature, given the heat flow:

$$q_1 = \frac{E_{b1} - J_1}{R_1} = \frac{\sigma \cdot T_1^4 - J_1}{R_1} \Rightarrow T_1 = \left[\frac{q_1 R_1 + J_1}{\sigma}\right]^{0.25}$$

Surface 2: Find the water wall heat input, given the water wall temperature:

$$q_2 = \frac{E_{b2} - J_2}{R_2} = \frac{\sigma \cdot T_2^4 - J_2}{R_2}$$

Surface 3: (Similar to surface 2) Find the water wall heat input, given the water wall temperature:

$$q_3 = \frac{E_{b3} - J_3}{R_3} = \frac{\sigma \cdot T_3^4 - J_3}{R_3}$$

BASIC CONCEPTS OF RADIATION FROM A SURFACE

Black-body Radiation

Blackbody radiation" or "cavity radiation" refers to an object or system which absorbs all radiation incident upon it and re-radiates energy which is characteristic of this radiating system only, not dependent upon the type of radiation which is incident upon it. The radiated energy can be considered to be produced by standing wave or resonant modes of the cavity which is radiating.

There was a general understanding of the mechanism involved — heat was known to cause the molecules and atoms of a solid to vibrate, and the molecules and atoms were themselves complicated patterns of electrical charges. (As usual, Newton was on the right track.) From the experiments of Hertz and others, Maxwell's predictions that oscillating charges emitted electromagnetic radiation had been confirmed, at least for simple antennas. It was known from Maxwell's equations that this radiation traveled at the speed of light and from this it was realized that light itself, and the closely related infrared heat radiation, were actually electromagnetic waves. The picture, then, was that when a body was heated, the consequent vibrations on a molecular and atomic scale inevitably induced charge oscillations. Assuming then that Maxwell's theory of electromagnetic radiation, which worked so well in the macroscopic world, was also valid at the molecular level, these oscillating charges would radiate, presumably giving off the heat and light observed.

How is Radiation Absorbed?

What is meant by the phrase "black body" radiation? The point is that the radiation from a heated body depends to some extent on the body being heated. To see this most easily, let's back up momentarily and consider how different materials absorb radiation. Some, like glass, seem to absorb light hardly at all—the light goes right through. For a shiny metallic surface, the light isn't absorbed either, it gets reflected. For a black material like soot, light and heat are almost completely absorbed, and the material gets warm. How can we understand these different behaviors in terms of light as an electromagnetic wave interacting with charges in the material, causing these charges to oscillate and absorb energy from the radiation? In the case of glass, evidently this doesn't happen, at least not much. Why not? A full understanding of why needs quantum mechanics, but the general idea is as follows: there are charges—electrons—in glass that are able to oscillate in response to an applied external oscillating electric field, but these charges are tightly bound to atoms, and can only oscillate at certain frequencies. (For quantum experts, these charge oscillations take place as an electron moves from one orbit to another. Of course, that was not understood in the 1890's, the time of the first precision work on black body radiation.) It happens that for ordinary glass none of these frequencies corresponds to visible light, so there is no resonance with a light wave, and hence little energy absorbed. That's why glass is perfect for windows! Duh. But glass is opaque at some frequencies outside the visible range (in general, both in the infrared and the ultraviolet). These are the frequencies at which the electrical charge distributions in the atoms or bonds can naturally oscillate.

How can we understand the reflection of light by a metal surface? A piece of metal has electrons free to move through the entire solid. This is what makes a metal a metal: it conducts both electricity and heat easily, both are actually carried by currents of these freely moving electrons. (Well, a little of the heat is carried by vibrations.) But metals are recognizable because they're shiny — why's that? Again, it's those free electrons: they're driven into large (relative to the atoms) oscillations by the electrical field of the incoming light wave, and this induced oscillating current radiates electromagnetically, just like a current in a transmitting antenna. This radiation is the reflected light. For a shiny metal surface, little of the incoming radiant energy is absorbed as heat, it's just reradiated, that is, reflected.

Now let's consider a substance that absorbs light: no transmission and no reflection. We come very close to perfect absorption with soot. Like a metal, it will conduct an electric current, but nowhere near as efficiently. There are unattached electrons, which can move through the whole solid, but they constantly bump into things — they have a short mean free path. When they bump, they cause vibration, like balls hitting bumpers in a pinball machine, so they give up kinetic energy into heat. Although the electrons in soot have a short mean free path compared to those in a good metal, they move very freely compared with electrons bound to atoms (as in glass), so they can accelerate and pick up energy from the electric field in the light wave. They are therefore very effective intermediaries in transferring energy from the light wave into heat.

Relating Absorption and Emission

Having seen how soot can absorb radiation and transfer the energy into heat, what about the reverse? Why does it radiate when heated? The pinball machine analogy is still good: imagine now a pinball machine where the barriers, etc., vibrate vigorously because they are being fed energy. The balls (the electrons) bouncing off them will be suddenly accelerated at each collision, and these accelerating charges emit electromagnetic waves. On the other hand, the electrons in a metal have very long mean free paths, the lattice vibrations affect them much less, so they are less effective in gathering and radiating away heat energy. It is evident from considerations like this that good absorbers of radiation are also good emitters.

In fact, we can be much more precise: a body emits radiation at a given temperature and frequency exactly as well as it absorbs the same radiation. This was proved by Kirchhoff: the essential point is that if we suppose a particular body can absorb better than it emits, then in a room full of objects all at the same temperature, it will absorb radiation from the other bodies better than it radiates energy back to them. This means it will get hotter, and the rest of the room will grow colder, contradicting the second law of thermodynamics. (We could use such a body to construct a heat engine extracting work as the room grows colder and colder).

But a metal glows when it's heated up enough: why is that? As the temperature is raised, the lattice of atoms vibrates more and more, these vibrations scatter and accelerate the electrons. Even glass glows at high enough temperatures, as the electrons are loosened and vibrate.

The "Black Body" Spectrum: A Hole in the Oven

Any body at any temperature above absolute zero will radiate to some extent, the intensity and frequency distribution of the radiation depending on the detailed structure of the body. To begin analyzing heat radiation, we need to be specific about the body doing the radiating: the simplest possible case is an idealized body which is a perfect absorber, and therefore also (from the above argument) a perfect emitter. For obvious reasons, this is called a "black body".

But we need to check our ideas experimentally: so how do we construct a perfect absorber? OK, nothing's perfect, but in 1859 Kirchhoff had a good idea: a small hole in the side of a large box is an excellent absorber, since any radiation that goes through the hole bounces around inside, a lot getting absorbed on each bounce, and has little chance of ever getting out again. So, we can do this in reverse: have an oven with a tiny hole in the side, and presumably the radiation coming out the hole is as good a representation of a perfect emitter as we're going to find. Kirchhoff challenged theorists and experimentalists to figure out and measure (respectively) the energy/frequency curve for this "cavity radiation", as he called it. In fact, it was Kirchhoff's challenge in 1859 that led directly to quantum theory forty years later.

What Was Observed: The Complete Picture

By the 1890's, experimental techniques had improved sufficiently that it was possible to make fairly precise measurements of the energy distribution in this cavity radiation, or as we shall call it black body radiation. In 1895, at the University of Berlin, Wien and

Lummer punched a small hole in the side of an otherwise completely closed oven, and began to measure the radiation coming out.

The beam coming out of the hole was passed through a diffraction grating, which sent the different wavelengths/frequencies in different directions, all towards a screen. A detector was moved up and down along the screen to find how much radiant energy was being emitted in each frequency range. (This is a theorist's model of the experiment—actual experimental arrangements were much more sophisticated. For example, to make the difficult infrared measurements higher frequency waves were eliminated by multiple reflections from quartz and other crystals.) They found a radiation intensity/frequency curve close to this (correct one):

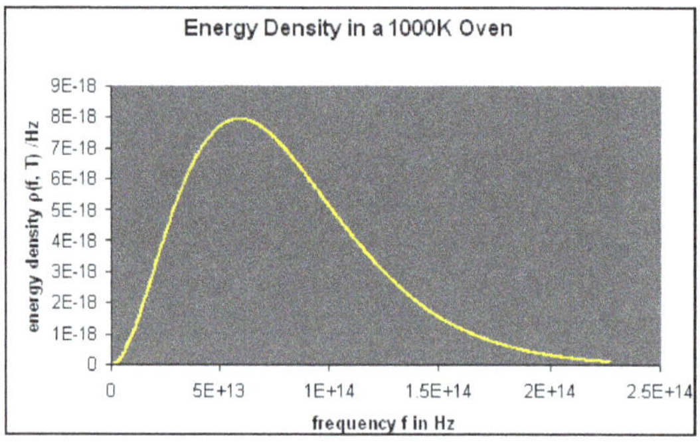

The visible spectrum begins at around 4.3×10^{14} Hz, so this oven glows deep red.

One minor point: this plot is the energy density inside the oven, which we denote by $\rho(f,T)$, meaning that at temperature T, the energy in Joules/m^3 in the frequency interval $f, f+\Delta f$ is $\rho(f,T)\,\Delta f$.

To find the power pumped out of the hole, bear in mind that the radiation inside the oven has waves equally going both ways — so only half of them will come out through the hole. Also, if the hole has area A, waves coming from the inside at an angle will see a smaller target area. The result of these two effects is that the radiation power from hole area,

$$A = \tfrac{1}{4} A c \rho(f,T).$$

They were also able to confirm both Stefan's Law $P = \sigma T^4$ and Wien's Displacement Law by measuring the black body curves at different temperatures, for example:

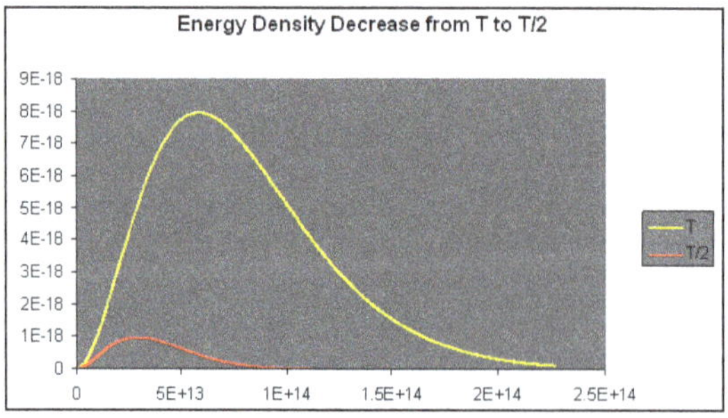

Let's look at these curves in more detail: for low frequencies f, $\rho(f,T)$ was found to be proportional to f^2, a parabolic shape, but for increasing f it fell below the parabola, peaking at f_{max}, then dropping quite rapidly towards zero as f increased beyond f_{max}.

For those low frequencies where $\rho(f,T)$ is parabolic, doubling the temperature was found to double the intensity of the radiation. But also at $2T$ the curve followed the doubled parabolic path much further before dropping away — in fact, twice as far, and $f_{max}(2T) = 2f_{max}(T)$.

The curve $\rho(f,2T)$ then, reaches eight times the height of $\rho(f,T)$. It also spreads over twice the lateral extent, so the area under the curve, corresponding to the total energy radiated, increases sixteen fold on doubling the temperature: Stefan's Law, $P = \sigma T^4$.

Understanding the Black Body Curve

These beautifully precise experimental results were the key to a revolution. The first successful theoretical analysis of the data was by Max Planck in 1900. He concentrated on modeling the oscillating charges that must exist in the oven walls, radiating heat inwards and — in thermodynamic equilibrium — themselves being driven by the radiation field.

The bottom line is that he found he could account for the observed curve if he required these oscillators not to radiate energy continuously, as the classical theory would demand, but they could only lose or gain energy in chunks, called quanta, of size hf, for an oscillator of frequency f. The constant h is now called Planck's constant, $h = 6.626 \times 10^{-34}$ *joule·sec.*

With that assumption, Planck calculated the following formula for the radiation energy density inside the oven:

$$\rho(f,T)df = \frac{8\pi V f^2 df}{c^3} \frac{hf}{e^{hf/kT} - 1}.$$

The perfect agreement of this formula with precise experiments, and the consequent necessity of energy quantization, was the most important advance in physics in the century.

But no-one noticed for several years! His black body curve was completely accepted as the correct one: more and more accurate experiments confirmed it time and again, yet the radical nature of the quantum assumption didn't sink in. Planck wasn't too upset — he didn't believe it either, he saw it as a technical fix that (he hoped) would eventually prove unnecessary.

Part of the problem was that Planck's route to the formula was long, difficult and implausible — he even made contradictory assumptions at different stages, as Einstein pointed out later. But the result was correct anyway, and to understand why we'll follow another, easier, route initiated (but not successfully completed) by Lord Rayleigh in England.

Rayleigh's Sound Idea: Counting Standing Waves

In 1900, actually some months before Planck's breakthrough work, Lord Rayleigh was taking a more direct approach to the radiation inside the oven: he didn't even think about oscillators in the walls, he just took the radiation to be a collection of standing waves in a cubical enclosure: electromagnetic oscillators. In contrast to the somewhat murky reality of the wall oscillators, these standing electromagnetic waves were crystal clear.

This was a natural approach for Rayleigh—he'd solved an almost identical problem a quarter century earlier, an analysis of standing sound waves in a cubical room. The task is to find and enumerate the different possible standing waves in the room/oven, compatible with the boundary conditions. For sound waves in a room, the amplitude of the sound goes to zero at the walls. For the electromagnetic waves, the electric field parallel to the wall must go to zero if the wall is a perfect conductor.

So what are the allowed standing waves? As a warm up exercise, consider the different allowed modes of vibration, that is, standing waves, in a string of length a fixed at both ends:

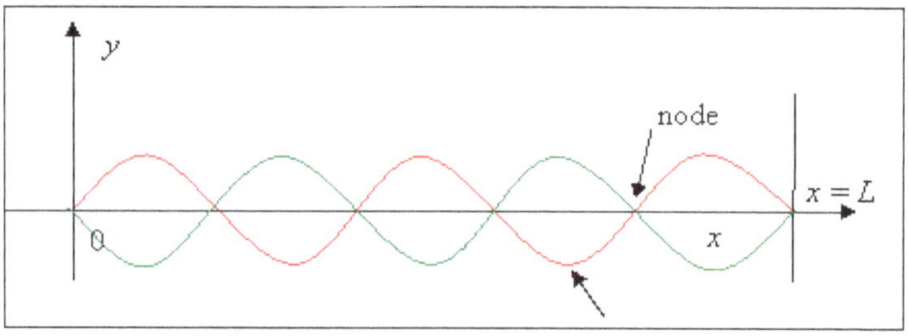

Possible mode of vibration of string with both ends fixed: $\lambda = 2L/5$.

The possible values of wavelength are:

$$\lambda = 2a,\ a,\ 2a/3,\ldots$$

So the allowed frequencies are:

$$f = c/\lambda = c/2a,\ 2(c/2a),\ 3(c/2a),\ldots$$

These allowed frequencies are equally spaced $c/2a$ apart. We define the spectral density by stating that number of modes between f and $f + \Delta f = N(f)\Delta f$.

Where we assume that Δf is large compared with the spacing between successive frequencies. Evidently for this one-dimensional exercise $N(f)$ is a constant equal to $2a/c$, each mode corresponds to an integer point on the real axis in units $c/2a$.

The amplitude of oscillation as a function of time is:

$$y = Asin\frac{2\pi x}{\lambda} sin\, 2\pi ft$$

more conveniently written,

$$y = Asin\, kx\, sin\omega t, where\ k = 2\pi / \lambda,\ \omega = 2\pi f, so\, \omega = ck.$$

The allowed values of k (called the wave number) are:

$$k = 2\pi / \lambda = \pi / a,\ 2\pi / a,\ 3\pi / a, \dots f = ck / 2\pi.$$

The generalization to three dimensions is simple: in a cubical box of side a, an allowed standing wave must satisfy the boundary conditions in all three directions. This means the choices of wave numbers are:

$$k_x = 2\pi / \lambda_x = \pi / a,\ 2\pi / a,\ 3\pi / a, \dots$$
$$k_y = 2\pi / \lambda_y = \pi / a,\ 2\pi / a,\ 3\pi / a,$$
$$k_z = 2\pi / \lambda_z = \pi / a,\ 2\pi / a,\ 3\pi / a, \dots$$

That is to say, each modes is labeled with three positive integers:

$$(k_x, k_y, k_z) = \frac{\pi}{a}(l, m, n)$$

and the frequency of the mode is:

$$f = ck / 2\pi = (c / 2\pi)\sqrt{k_x^2 + k_y^2 + k_z^2}.$$

For infrared and visible radiation in a reasonable sized oven, frequency intervals measured experimentally are far greater than the spacing $c / 2a$ of these integer points. Just as in the one-dimensional example, these modes fill the three-dimensional k-space uniformly, with density $(a / \pi)^3$, but now this means the mode density is not uniform as a function of frequency.

The number of them between f and $f + \Delta f = N(f)\Delta f$ is the volume in k - space, in units $(a / \pi)^3$, of the spherical shell of radius $k = 2\pi f / c$, thickness $\Delta k = 2\pi \Delta f / c$, and restricted to all components of k being positive (like the integers), a factor of 1/8.

Including a factor of 2 for the two polarization states of the standing electromagnetic waves, the density of states as a function of frequency in an oven of volume $V = a^3$ is:

$$N(f)\Delta f = \frac{1}{8} \times 2 \times \frac{4\pi k^2 \Delta k}{(\pi / a)^3} = \frac{1}{4} \times \left(\frac{a}{\pi}\right)^3 \times 4\pi \left(\frac{2\pi}{c}\right)^3 f^2 \Delta f$$

What about Equipartition of Energy?

A central result of classical statistical mechanics is the equipartition of energy: for a system in thermal equilibrium, each degree of freedom has average energy $\frac{1}{2}k_{\mathrm{B}}T$. k_B being Boltzmann's constant. Thus molecules in a gas have average kinetic energy $\frac{3}{2}k_BT$, $\frac{1}{2}k_BT$, for each direction, and a simple one-dimensional harmonic oscillator has total energy k_BT : $\frac{1}{2}k_BT$ kinetic energy and $\frac{1}{2}k_{\mathrm{B}}T$ potential energy.

Comparing now the formula for the number of modes $N(f)\Delta f$ in a small interval Δf,

$$N(f)\Delta f = \frac{8V\pi f^2\Delta f}{c^3}$$

with Planck's formula for radiation energy intensity in the same interval:

$$\rho(f,T)\Delta f = \frac{8\pi V f^2\Delta f}{c^3}\frac{hf}{e^{hf/k_BT}-1},$$

for the low frequency modes $hf \ll k_{\mathrm{B}}T$ we can make the approximation,

$$e^{hf/k_BT} - 1 \cong hf/k_BT$$

and it follows immediately that each mode has energy k_BT,, in line with classical predictions.

But things go badly wrong at high frequencies! The number of modes increases without limit, the energy in these high frequency modes, though, is decaying exponentially as the frequency increases. Ehrenfest later dubbed this the ultraviolet catastrophe. Rayleigh's sound approach apparently wasn't so sound after all — something crucial was missing.

It is perhaps surprising that Planck never mentioned equipartition. Of course, as Rayleigh himself remarked, equipartition was well-known to have problems, for example in the specific heat of gases. And in fact Planck wasn't even sure about the existence of atoms: he later wrote that in the 1890's "I had been inclined to reject atomism". In fact, even Boltzmann was very unsure how well oscillators came to thermal equilibrium with electromagnetic radiation — after all, it was well known that oscillation of diatomic molecules failed to reach classical thermal equilibrium with kinetic energy. (As long ago as 1877, Maxwell had pointed out that hot gases emit light at particular frequencies. The frequencies do not change with temperature, so the oscillations must be simple harmonic — but such an oscillator would surely also be excited by collisions at low temperatures, so why was energy not being fed into this mode).

Einstein Sees a Gas of Photons

After Planck announced his result in December, 1900 there was a deafening silence on the subject for several years. No-one (including Planck) realized the importance of what he had done — his work was widely seen as just a clever technical fix, even if it did give the right answer (the curve itself was completely accepted as correct).

Then in March, 1905, Albert Einstein turned his attention to the problem. He first rederived the Rayleigh result assuming equipartition:

$$\rho(f,T)=\frac{8\pi f^2}{c^3}k_BT$$

and observed that this made no sense at high frequencies. So he focused on Planck's formula for high frequencies, $hf \gg k_BT$:

$$\rho(f,T)df \cong \frac{8\pi Vhf^3 df}{c^3}e^{-hf/k_BT}$$

(asymptotically identical for $f \to \infty$ to an earlier formula by Wien).

Einstein perceived an analogy here with the energy distribution in a classical gas.

The (normalized) probability distribution function for classical atoms as a function of speed was,

$$f(v)=4\pi\left(\frac{m}{2\pi kT}\right)^{3/2}v^2e^{-E/k_BT}$$

and the corresponding energy density in E is,

$$f(E)=\left[4\pi\left(\frac{m}{2\pi kT}\right)^{3/2}v^2\right]Ee^{-E/k_BT}.$$

The radiation formula at high frequencies is,

$$\rho(f,T)=\frac{8\pi f^2hf}{c^3}e^{-hf/k_BT}.$$

Einstein pointed out that if the high frequency radiation is imagined to be a gas of independent particles having energy E=hf, the energy density in frequency/energy in the radiation is,

$$\rho(E,T)=\left[\frac{8\pi f^2}{c^3}\right]Ee^{-E/k_BT}.$$

Comparing this with the expression for atoms, the analogy is close: recall that for the radiation, frequency is proportional to wave number and, on quantization, to momentum; for the (nonrelativistic) atoms velocity is proportional to momentum, so both these distributions are essentially in momentum space. Of course, the normalization factors differ, because the total number of atoms doesn't change with temperature, unlike the total radiation. Nevertheless, the analogy is compelling, and led Einstein to state that the radiation in the enclosure was itself quantized, the energy quantization was not some special property only of the wall oscillators, as Planck thought. The radiation quanta are of course photons, but that word wasn't coined until later.

Einstein had been troubled by Planck's derivation of his result, depending as it did first, on a classical analysis of the interaction between the wall oscillator and radiation, followed by a claim that the interaction was in fact not like that at all. But the answer was right, and now Einstein began to see why. In contrast to the poorly understood wall oscillators, the electromagnetic standing wave oscillations in the oven were completely clear.

Energy in an Oscillator as a Function of Temperature

Einstein realized that, in terms of Rayleigh's electromagnetic standing waves, the blackbody radiation curves have a simple interpretation: the average energy in an oscillator of frequency f at temperature T is,

$$\overline{E} = \frac{hf}{e^{hf/k_B T} - 1}.$$

Furthermore, Planck's work made plausible that this same quantization held for the material oscillators in the walls.

Einstein took the next step: he conjectured that all oscillators are quantized, for example a vibrating atom in a solid. This would explain why the Dulong Petit law, which assigns specific heat $3k_B$ to each atom in a solid, does not hold good at low temperatures: once $k_B T \ll hf$, the modes are not excited, so absorb little heat. The specific heat falls, as is indeed observed. Furthermore, it explains why diatomic gas molecules, such as oxygen and nitrogen, do not appear to absorb heat into vibrational modes—these modes have very high frequency.

It's worth thinking about the constant exchange of energy with the environment for an oscillator in thermal equilibrium at temperature T. The random thermal fluctuations in a system have energy of order $k_B T$, this is the amount of energy, approximately, delivered back and forth. But if an oscillator has $hf=5k_B T$, say, it can only accept chunks of energy of size $5k_B T$, and will only be excited in the unlikely event that five of these random $k_B T$ fluctuations come together at the right place at the right time. The high frequency modes are effectively frozen out by this minimum energy requirement. The exponential drop off in excitation with frequency reflects the exponential drop off in probability of getting the right number of fluctuations together, analogous to the exponential drop off in probability of tossing a coin n heads in a row.

Simple Derivation of Planck's Formula from the Boltzmann's Distribution

Planck's essential assumption in deriving his formula was that the oscillators only exchange energy with the radiation in quanta hf. Einstein made clear that the well-understood standing electromagnetic waves, the radiation in the oven, also have quantized energies.

The probability of a system at temperature T having energy E is proportional to $e{-}^{E/kBT}$, Boltzmann's formula. It turns out that this formula continues to be valid in quantum systems. Now, a classical simple harmonic oscillator at T will have a probability distribution proportional to $e^{-E/kBT} = e^{-(mv2+m\omega 2x2)/2kBT}$, so the expectation value of the energy is,

$$\overline{E} = \frac{\iint \left(\frac{1}{2}mv^2 + \frac{1}{2}m\omega^2 x^2 \right) e^{-(mv^2+m\omega^2x^2)/2k_BT} dvdx}{\iint e^{-(mv^2+m\omega^2x^2)/2k_BT} dvdx} = k_B T,$$

just the classical equipartition of energy.

But we now know this isn't true if the oscillator is quantized: the energies are now in steps hf apart. Taking the ground state as the zero of energy, allowed energies are

$$0, hf, 2hf, 3hf \ldots$$

and assuming the Boltzmann expression for relative probabilities is still correct, the relative probabilities of these states will be in the ratios:

$$e^{-hf/k_BT}, e^{-2hf/k_BT}, e^{-3hf/k_BT} \ldots$$

To find the oscillator energy at this temperature, we use these probabilities weighted by the corresponding energy, and divide by a normalization factor to ensure that the probabilities add up to 1:

$$\overline{E} = \frac{hfe^{-hf/k_BT} + 2hfe^{-2hf/k_BT} + 3hfe^{-3hf/k_BT} \ldots}{1 + e^{-hf/k_BT} + e^{-2hf/k_BT} + e^{-3hf/k_BT} \ldots}$$
$$= \frac{hf}{e^{hf/k_BT} - 1}.$$

(The expression is evaluated as follows: write $e^{-hf/kT}=x$, so the sum of the relative probabilities is $1 + x + x^2 + x^3 + \ldots = 1/(1-x)$ and the numerator in the above expression for

$\bar{E}$ is $hfx(1+2x+3x^2+\ldots)=hfx/(1-x)^2$, since the infinite series in parens is given by differentiating 1 1+x+x2+x3+...)

This is indeed the correct result from the black body experiments. Evidently Boltzmann's relative probability function e−E/kBT is still valid in quantum systems.

Planck's Law

Planck's law describes the spectral density of electromagnetic radiation emitted by a black body in thermal equilibrium at a given temperature *T*, when there is no net flow of matter or energy between the body and its environment.

At the end of the 19th century, physicists were unable to explain why the observed spectrum of black body radiation, which by then had been accurately measured, diverged significantly at higher frequencies from that predicted by existing theories. In 1900, Max Planck heuristically derived a formula for the observed spectrum by assuming that a hypothetical electrically charged oscillator in a cavity that contained black-body radiation could only change its energy in a minimal increment, *E*, that was proportional to the frequency of its associated electromagnetic wave. This resolved the problem of the ultraviolet catastrophe predicted by classical physics.

It was a pioneering insight of modern physics and is of fundamental importance to quantum theory.

The Law

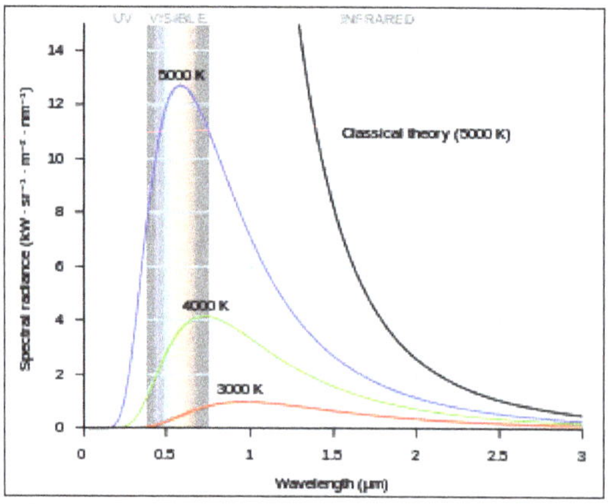

Planck's law accurately describes black-body radiation.

Shown here are a family of curves for different temperatures. The classical (black) curve diverges from observed intensity at high frequencies.

Every physical body spontaneously and continuously emits electromagnetic radiation and the spectral radiance of a body, B_ν, describes the amount of energy it emits at different radiation frequencies. It is the power emitted per unit area of the body, per unit solid angle of emission, per unit frequency. Planck showed that the spectral radiance of a body for frequency v at absolute temperature T is given by,

$$B_v(v,T) = \frac{2hv^3}{c^2} \frac{1}{e^{\frac{hv}{k_B T}} - 1}$$

where k_B is the Boltzmann constant, h is the Planck constant, and c is the speed of light in the medium, whether material or vacuum. The spectral radiance can also be expressed per unit wavelength λ instead of per unit frequency. In this case, it is given by.

$$B_\lambda(\lambda,T) = \frac{2hc^2}{\lambda^5} \frac{1}{e^{\frac{hc}{\lambda k_B T}} - 1},$$

showing how radiated energy emitted at shorter wavelengths increases more rapidly with temperature than energy emitted at longer wavelengths. The law may also be expressed in other terms, such as the number of photons emitted at a certain wavelength, or the energy density in a volume of radiation. The SI units of B_v are $W \cdot sr^{-1} \cdot m^{-2} \cdot Hz^{-1}$, while those of B_λ are $W \cdot sr^{-1} \cdot m^{-3}$.

In the limit of low frequencies (i.e. long wavelengths), Planck's law tends to the Rayleigh–Jeans law, while in the limit of high frequencies (i.e. small wavelengths) it tends to the Wien approximation.

Max Planck developed the law in 1900 with only empirically determined constants, and later showed that, expressed as an energy distribution, it is the unique stable distribution for radiation in thermodynamic equilibrium. As an energy distribution, it is one of a family of thermal equilibrium distributions which include the Bose–Einstein distribution, the Fermi–Dirac distribution and the Maxwell–Boltzmann distribution.

Different Forms

Planck's law can be encountered in several forms depending on the conventions and preferences of different scientific fields. The various forms of the law for spectral radiance are summarized in the table below. Forms on the left are most often encountered in experimental fields, while those on the right are most often encountered in theoretical fields.

Planck's law expressed in terms of different spectral variables			
with h		with $\hbar$	
variable	distribution	variable	distribution
Frequency v	$B_v(v,T) = \frac{2hv^3}{c^2} \frac{1}{e^{hv/(k_B T)} - 1}$	Angular frequency ω	$B_\omega(\omega,T) = \frac{\hbar\omega^3}{4\pi^3 c^2} \frac{1}{e^{\hbar\omega/(k_B T)} - 1}$

Wavelength λ	$B_\lambda(\lambda,T)=\frac{2hc^2}{\lambda^5}\frac{1}{e^{hc/(\lambda k_B T)}-1}$	Angular wavelength y	$B_y(y,T)=\frac{\hbar c^2}{4\pi^3 y^5}\frac{1}{e^{\hbar c/(y k_B T)}-1}$
Wavenumber $\tilde{\nu}$	$B_{\tilde{\nu}}(\tilde{\nu},T)=2hc^2\tilde{\nu}^3\frac{1}{e^{hc\tilde{\nu}/(k_B T)}-1}$	Angular wavenumber k	$B_k(k,T)=\frac{\hbar c^2 k^3}{4\pi^3}\frac{1}{e^{\hbar ck/(k_B T)}-1}$

These distributions represent the spectral radiance of blackbodies—the power emitted from the emitting surface, per unit projected area of emitting surface, per unit solid angle, per spectral unit (frequency, wavelength, wavenumber or their angular equivalents). Since the radiance is isotropic (i.e. independent of direction), the power emitted at an angle to the normal is proportional to the projected area, and therefore to the cosine of that angle as per Lambert's cosine law, and is unpolarized.

Correspondence between Spectral Variable forms

Different spectral variables require different corresponding forms of expression of the law. In general, one may not convert between the various forms of Planck's law simply by substituting one variable for another, because this would not take into account that the different forms have different units. Wavelength and frequency units are reciprocal.

Corresponding forms of expression are related because they express one and the same physical fact: for a particular physical spectral increment, a corresponding particular physical energy increment is radiated.

This is so whether it is expressed in terms of an increment of frequency, dν, or, correspondingly, of wavelength, dλ. Introduction of a minus sign can indicate that an increment of frequency corresponds with decrement of wavelength. In order to convert the corresponding forms so that they express the same quantity in the same units we multiply by the spectral increment. Then, for a particular spectral increment, the particular physical energy increment may be written,

$B_\lambda(\lambda,T)d\lambda=-B_\nu(\nu(\lambda),T)d\nu$, which leads to $B_\lambda(\lambda,T)=-\frac{d\nu}{d\lambda}B_\nu(\nu(\lambda),T)$.

Also, $\nu(\lambda)=\frac{c}{\lambda}$, so that $\frac{d\nu}{d\lambda}=-\frac{c}{\lambda^2}$. Substitution gives the correspondence between the frequency and wavelength forms, with their different dimensions and units. Consequently,

$$\frac{B_\lambda(T)}{B_\nu(T)}=\frac{c}{\lambda^2}=\frac{\nu^2}{c}.$$

Evidently, the location of the peak of the spectral distribution for Planck's law depends on the choice of spectral variable. Nevertheless, in a manner of speaking, this formula means that the shape of the spectral distribution is independent of temperature.

Spectral Energy Density Form

Planck's law can also be written in terms of the spectral energy density (u) by multiplying B by $\frac{4\pi}{c}$.

$$u_i(T) = \frac{4\pi}{c} B_i(T).$$

These distributions have units of energy per volume per spectral unit.

First and Second Radiation Constants

In the above variants of Planck's law, the *Wavelength* and *Wavenumber* variants use the terms $2hc^2$ and $\frac{hc}{k_B}$ which comprise physical constants only. Consequently, these terms can be considered as physical constants themselves, and are therefore referred to as the first radiation constant c_{1L} and the second radiation constant c_2 with,

$$c_{1L} = 2hc^2$$

and

$$c_2 = \frac{hc}{k_B}$$

Using the radiation constants, the *Wavelength* variant of Planck's law can be simplified to,

$$L(\lambda,T) = \frac{c_{1L}}{\lambda^5} \frac{1}{\exp\left(\frac{c_2}{\lambda T}\right) - 1}.$$

and the *wavenumber* variant can be simplified correspondingly.

L is used here instead of B because it is the SI symbol for *spectral radiance*. The L in c_{1L} refers to that. This reference is necessary because Planck's law can be reformulated to give spectral radiant exitance $M(\lambda,T)$ rather than *spectral radiance* $L(\lambda,T)$, in which case c_1 replaces c_{1L}, with

$$c_1 = 2\pi hc^2,$$

so that Planck's law for *spectral radiant exitance* can be written as,

$$M(\lambda,T) = \frac{c_1}{\lambda^5} \frac{1}{\exp\left(\frac{c_2}{\lambda T}\right) - 1}$$

Physics

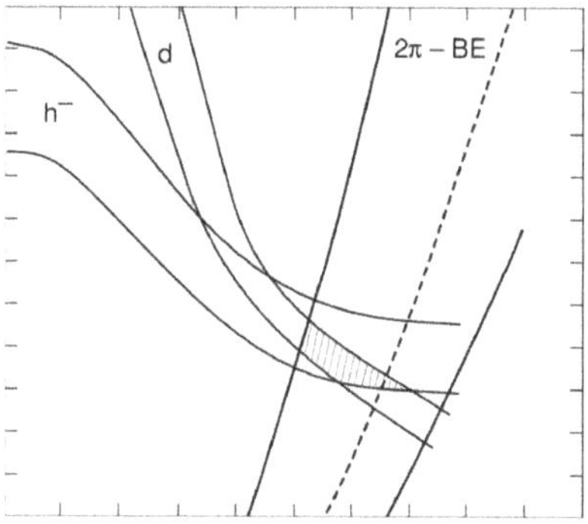

Freeze-out of high-energy oscillators.

Planck's law describes the unique and characteristic spectral distribution for electromagnetic radiation in thermodynamic equilibrium, when there is no net flow of matter or energy. Its physics is most easily understood by considering the radiation in a cavity with rigid opaque walls. Motion of the walls can affect the radiation. If the walls are not opaque, then the thermodynamic equilibrium is not isolated. It is of interest to explain how the thermodynamic equilibrium is attained. There are two main cases: (a) when the approach to thermodynamic equilibrium is in the presence of matter, when the walls of the cavity are imperfectly reflective for every wavelength or when the walls are perfectly reflective while the cavity contains a small black body (this was the main case considered by Planck); or (b) when the approach to equilibrium is in the absence of matter, when the walls are perfectly reflective for all wavelengths and the cavity contains no matter. For matter not enclosed in such a cavity, thermal radiation can be approximately explained by appropriate use of Planck's law.

Classical physics led, via the equipartition theorem, to the ultraviolet catastrophe, a prediction that the total blackbody radiation intensity was infinite. If supplemented by the classically unjustifiable assumption that for some reason the radiation is finite, classical thermodynamics provides an account of some aspects of the Planck distribution, such as the Stefan–Boltzmann law, and the Wien displacement law. For the case of the presence of matter, quantum mechanics provides a good account, as found below in the section headed Einstein coefficients. This was the case considered by Einstein, and is nowadays used for quantum optics. For the case of the absence of matter, quantum field theory is necessary, because non-relativistic quantum mechanics with fixed particle numbers does not provide a sufficient account.

Photons

Quantum theoretical explanation of Planck's law views the radiation as a gas of massless, uncharged, bosonic particles, namely photons, in thermodynamic equilibrium. Photons are viewed as the carriers of the electromagnetic interaction between electrically charged elementary particles. Photon numbers are not conserved. Photons are created or annihilated in the right numbers and with the right energies to fill the cavity with the Planck distribution. For a photon gas in thermodynamic equilibrium, the internal energy density is entirely determined by the temperature; moreover, the pressure is entirely determined by the internal energy density. This is unlike the case of thermodynamic equilibrium for material gases, for which the internal energy is determined not only by the temperature, but also, independently, by the respective numbers of the different molecules, and independently again, by the specific characteristics of the different molecules. For different material gases at given temperature, the pressure and internal energy density can vary independently, because different molecules can carry independently different excitation energies.

Planck's law arises as a limit of the Bose–Einstein distribution, the energy distribution describing non-interactive bosons in thermodynamic equilibrium. In the case of massless bosons such as photons and gluons, the chemical potential is zero and the Bose–Einstein distribution reduces to the Planck distribution. There is another fundamental equilibrium energy distribution: the Fermi–Dirac distribution, which describes fermions, such as electrons, in thermal equilibrium. The two distributions differ because multiple bosons can occupy the same quantum state, while multiple fermions cannot. At low densities, the number of available quantum states per particle is large, and this difference becomes irrelevant. In the low density limit, the Bose–Einstein and the Fermi–Dirac distribution each reduce to the Maxwell–Boltzmann distribution.

Radiative Transfer

The equation of radiative transfer describes the way in which radiation is affected as it travels through a material medium. For the special case in which the material medium is in thermodynamic equilibrium in the neighborhood of a point in the medium, Planck's law is of special importance.

For simplicity, we can consider the linear steady state, without scattering. The equation of radiative transfer states that for a beam of light going through a small distance ds, energy is conserved: The change in the (spectral) radiance of that beam (I_ν) is equal to the amount removed by the material medium plus the amount gained from the material medium. If the radiation field is in equilibrium with the material medium, these two contributions will be equal. The material medium will have a certain emission coefficient and absorption coefficient.

The absorption coefficient α is the fractional change in the intensity of the light beam as it travels the distance ds, and has units of length^{-1}. It is composed of two parts, the decrease due to absorption and the increase due to stimulated emission. Stimulated emission is emission by the material body which is caused by and is proportional to the incoming radiation. It is included in the absorption term because, like absorption, it is proportional to the intensity of the incoming radiation. Since the amount of absorption will generally vary linearly as the density ρ of the material, we may define a "mass absorption coefficient" $\kappa_\nu = \frac{\alpha}{\rho}$ which is a property of the material itself. The change in intensity of a light beam due to absorption as it traverses a small distance d*s* will then be

$$dI_\nu = -\kappa_\nu \rho I_\nu ds$$

The "mass emission coefficient" j_ν is equal to the radiance per unit volume of a small volume element divided by its mass (since, as for the mass absorption coefficient, the emission is proportional to the emitting mass) and has units of power·solid angle^{-1}·frequency^{-1}·density^{-1}. Like the mass absorption coefficient, it too is a property of the material itself. The change in a light beam as it traverses a small distance d*s* will then be

$$dI_\nu = j_\nu \rho ds$$

The equation of radiative transfer will then be the sum of these two contributions:

$$\frac{dI_\nu}{ds} = j_\nu \rho - \kappa_\nu \rho I_\nu.$$

If the radiation field is in equilibrium with the material medium, then the radiation will be homogeneous (independent of position) so that $dI_\nu = 0$ and:

$$\kappa_\nu B_\nu = j_\nu$$

which is another statement of Kirchhoff's law, relating two material properties of the medium, and which yields the radiative transfer equation at a point around which the medium is in thermodynamic equilibrium:

$$\frac{dI_\nu}{ds} = \kappa_\nu \rho (B_\nu - I_\nu).$$

Einstein Coefficients

The principle of detailed balance states that, at thermodynamic equilibrium, each elementary process is equilibrated by its reverse process.

In 1916, Albert Einstein applied this principle on an atomic level to the case of an atom radiating and absorbing radiation due to transitions between two particular energy levels, giving a deeper insight into the equation of radiative transfer and Kirchhoff's law for this type of radiation. If level 1 is the lower energy level with energy E_1, and level 2 is the upper energy level with energy E_2, then the frequency ν of the radiation radiated or absorbed will be determined by Bohr's frequency condition:

$$E_2 - E_1 = h\nu.$$

If n_1 and n_2 are the number densities of the atom in states 1 and 2 respectively, then the rate of change of these densities in time will be due to three processes:

$$\left(\frac{dn_1}{dt}\right)_{\text{spon}} = A_{21}n_2u_v \text{ Spontaneous emission}$$

$$\left(\frac{dn_1}{dt}\right)_{\text{stim}} = B_{21}n_2u_v \text{ Stimulated emission}$$

$$\left(\frac{dn_2}{dt}\right)_{\text{abs}} = B_{12}n_1u_v \text{ Photo-absorption}$$

where u_v is the spectral energy density of the radiation field. The three parameters A_{21}, B_{21} and B_{12}, known as the Einstein coefficients, are associated with the photon frequency v produced by the transition between two energy levels (states). As a result, each line in a spectra has its own set of associated coefficients. When the atoms and the radiation field are in equilibrium, the radiance will be given by Planck's law and, by the principle of detailed balance, the sum of these rates must be zero:

$$0 = A_{21}n_2 + B_{21}n_2\frac{4\pi}{c}B_v(T) - B_{12}n_1\frac{4\pi}{c}B_v(T)$$

Since the atoms are also in equilibrium, the populations of the two levels are related by the Boltzmann factor:

$$\frac{n_2}{n_1} = \frac{g_2}{g_1}e^{-hv/k_B T}$$

where g_1 and g_2 are the multiplicities of the respective energy levels. Combining the above two equations with the requirement that they be valid at any temperature yields two relationships between the Einstein coefficients:

$$\frac{A_{21}}{B_{21}} = \frac{8\pi hv^3}{c^3}$$

$$\frac{B_{21}}{B_{12}} = \frac{g_1}{g_{2\,v}}.$$

so that knowledge of one coefficient will yield the other two. For the case of isotropic absorption and emission, the emission coefficient (j_v) and absorption coefficient (κ_v) defined in the radiative transfer section above, can be expressed in terms of the Einstein coefficients. The relationships between the Einstein coefficients will yield the expression of Kirchhoff's law,namely that,

$$j_v = \kappa_v B_v$$

These coefficients apply to both atoms and molecules.

Properties

Peaks

The distributions B_ν, B_ω, $B_{\tilde{\nu}}$ and B_k peak at a photon energy of,

$$E = \left[3 + W\left(\frac{-3}{e^3}\right)\right] k_B T \approx 2.821\, k_B T,$$

where, W is the Lambert W function and e is Euler's number.

The distributions B_λ and B_y however, peak at a different energy,

$$E = \left[5 + W\left(\frac{-5}{e^5}\right)\right] k_B T \approx 4.965\, k_B T,$$

The reason for this is that, as mentioned above, one cannot go from (for example) B_ν to B_λ simply by substituting ν by λ. In addition, one must also multiply the result of the substitution by.

$$\left|\frac{d\nu}{d\lambda}\right| = c / \lambda^2.$$

This $\frac{1}{\lambda^2}$ factor shifts the peak of the distribution to higher energies. These peaks are the *mode* energy of a photon, when binned using equal-size bins of frequency or wavelength, respectively. Meanwhile, the *average* energy of a photon from a blackbody is,

$$E = \left[360 \frac{\zeta(5)}{\pi^4}\right] k_B T \approx 3.832\, k_B T,$$

where, ζ is the Riemann zeta function. Dividing hc by this energy expression gives the wavelength of the peak. For this one can use $\frac{hc}{k_B} = 14387.770$ μm·K.

The spectral radiance at these peaks is given by:

$$B_{\nu,\max}(T) = \frac{2k_B^3 T^3 (3 + W(-3\exp(-3)))^3}{h^2 c^2} \frac{1}{e^{3+W(-3\exp(-3))} - 1} \approx \left(1.896 \times 10^{-19} \frac{\text{W}}{\text{m}^2 \cdot \text{Hz} \cdot \text{K}^3}\right) \times T^3$$

$$B_{\lambda,\max}(T) = \frac{2k_B^5 T^5 (5 + W(-5\exp(-5)))^5}{h^4 c^3} \frac{1}{e^{5+W(-5\exp(-5))} - 1} \approx \left(4.096 \times 10^{-6} \frac{\text{W}}{\text{m}^3 \cdot \text{K}^5}\right) \times T^5$$

Approximations

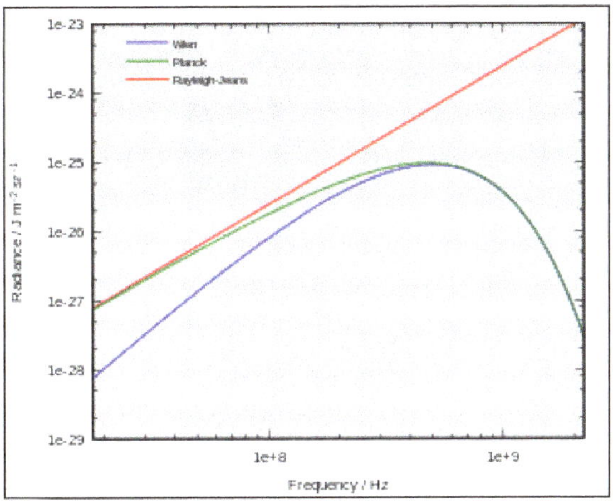

Log-log plots of radiance vs. frequency for Planck's law (green), compared with the Rayleigh–Jeans law (red) and the Wien approximation (blue) for a black body at 8 mK temperature.

In the limit of low frequencies (i.e. long wavelengths), Planck's law becomes the Rayleigh–Jeans law,

$$B_\nu(T) \approx \frac{2\nu^2}{c^2} k_B T \text{ or } B_\lambda(T) \approx \frac{2c}{\lambda^4} k_B T.$$

The radiance increases as the square of the frequency, illustrating the ultraviolet catastrophe. In the limit of high frequencies (i.e. small wavelengths) Planck's law tends to the Wien approximation:

$$B_\nu(T) \approx \frac{2h\nu^3}{c^2} e^{-\frac{h\nu}{k_B T}} \text{ or } B_\lambda(T) \approx \frac{2hc^2}{\lambda^5} e^{-\frac{hc}{\lambda k_B T}}.$$

Both approximations were known to Planck before he developed his law. He was led by these two approximations to develop a law which incorporated both limits, which ultimately became Planck's law.

Percentiles

Wien's displacement law in its stronger form states that the shape of Planck's law is independent of temperature. It is therefore possible to list the percentile points of the total radiation as well as the peaks for wavelength and frequency, in a form which gives the wavelength λ when divided by temperature T. The second row of the following table lists the corresponding values of λT, that is, those values of x for which the wavelength λ is $\frac{x}{T}$ micrometers at the radiance percentile point given by the corresponding entry in the first row.

Percentile	0.01%	0.1%	1%	10%	20%	25.0%	30%	40%	41.8%
λT (μm·K)	910	1110	1448	2195	2676	2898	3119	3582	3670

Percentile	50%	60%	64.6%	70%	80%	90%	99%	99.9%	99.99%
λT (μm·K)	4107	4745	5099	5590	6864	9376	22884	51613	113374

That is, 0.01% of the radiation is at a wavelength below $\frac{910}{T}$ μm, 20% below $\frac{2676}{T}$ μm, etc. The wavelength and frequency peaks are in bold and occur at 25.0% and 64.6% respectively. The 41.8% point is the wavelength-frequency-neutral peak. These are the points at which the respective Planck-law functions $\frac{1}{\lambda^5}$, $v^3 and \frac{v^2}{\lambda^2}$ divided by exp $(\frac{hv}{k_B T})-1$ attain their maxima. The much smaller gap in ratio of wavelengths between 0.1% and 0.01% (1110 is 22% more than 910) than between 99.9% and 99.99% (113374 is 120% more than 51613) reflects the exponential decay of energy at short wavelengths (left end) and polynomial decay at long.

Which peak to use depends on the application. The conventional choice is the wavelength peak at 25.0% given by Wien's displacement law in its weak form. For some purposes the median or 50% point dividing the total radiation into two halves may be more suitable. The latter is closer to the frequency peak than to the wavelength peak because the radiance drops exponentially at short wavelengths and only polynomially at long. The neutral peak occurs at a shorter wavelength than the median for the same reason.

For the Sun, *T* is 5778 K, allowing the percentile points of the Sun's radiation, in nanometers, to be tabulated as follows when modeled as a black body radiator, to which the Sun is a fair approximation. For comparison a planet modeled as a black body radiating at a nominal 288 K (15 °C) as a representative value of the Earth's highly variable temperature has wavelengths more than twenty times that of the Sun, tabulated in the third row in micrometers (thousands of nanometers).

Percentile	0.01%	0.1%	1%	10%	20%	25.0%	30%	40%	41.8%
Sun λ (μm)	0.157	0.192	0.251	0.380	0.463	0.502	0.540	0.620	0.635
288 K planet λ (μm)	3.16	3.85	5.03	7.62	9.29	10.1	10.8	12.4	12.7

Percentile	50%	60%	64.6%	70%	80%	90%	99%	99.9%	99.99%
Sun λ (μm)	0.711	0.821	0.882	0.967	1.188	1.623	3.961	8.933	19.620
288 K planet λ (μm)	14.3	16.5	17.7	19.4	23.8	32.6	79.5	179	394

That is, only 1% of the Sun's radiation is at wavelengths shorter than 251 nm, and only 1% at longer than 3961 nm. Expressed in micrometers this puts 98% of the Sun's radiation in the range from 0.251 to 3.961 μm. The corresponding 98% of energy radiated from a 288 K planet is from 5.03 to 79.5 μm, well above the range of solar radiation (or below if expressed in terms of frequencies $\nu = c/\lambda$ instead of wavelengths λ).

A consequence of this more-than-order-of-magnitude difference in wavelength between solar and planetary radiation is that filters designed to pass one and block the other are easy to construct. For example, windows fabricated of ordinary glass or transparent plastic pass at least 80% of the incoming 5778 K solar radiation, which is below 1.2 μm in wavelength, while blocking over 99% of the outgoing 288 K thermal radiation from 5 μm upwards, wavelengths at which most kinds of glass and plastic of construction-grade thickness are effectively opaque.

The Sun's radiation is that arriving at the top of the atmosphere (TOA). As can be read from the table, radiation below 400 nm, or ultraviolet, is about 12%, while that above 700 nm, or infrared, starts at about the 49% point and so accounts for 51% of the total. Hence only 37% of the TOA insolation is visible to the human eye. The atmosphere shifts these percentages substantially in favor of visible light as it absorbs most of the ultraviolet and significant amounts of infrared.

Derivation

Consider a cube of side L with conducting walls filled with electromagnetic radiation in thermal equilibrium at temperature T. If there is a small hole in one of the walls, the radiation emitted from the hole will be characteristic of a perfect black body. We will first calculate the spectral energy density within the cavity and then determine the spectral radiance of the emitted radiation.

At the walls of the cube, the parallel component of the electric field and the orthogonal component of the magnetic field must vanish. Analogous to the wave function of a particle in a box, one finds that the fields are superpositions of periodic functions. The three wavelengths λ_1, λ_2, and λ_3, in the three directions orthogonal to the walls can be:

$$\lambda_i = \frac{2L}{n_i},$$

where the n_i are positive integers. For each set of integers n_i there are two linearly independent solutions (known as modes). According to quantum theory, the energy levels of a mode are given by:

$$E_{n_1,n_2,n_3}(r) = \left(r + \frac{1}{2}\right)\frac{hc}{2L}\sqrt{n_1^2 + n_2^2 + n_3^2}.$$

The quantum number r can be interpreted as the number of photons in the mode. The two modes for each set of n_i correspond to the two polarization states of the photon which has a spin of 1. For $r = 0$ the energy of the mode is not zero. This vacuum energy of the electromagnetic field is responsible for the Casimir effect. In the following we will calculate the internal energy of the box at absolute temperature T.

According to statistical mechanics, the equilibrium probability distribution over the energy levels of a particular mode is given by:

$$P_r = \frac{\exp(-\beta E(r))}{Z(\beta)}.$$

Here,

$$\beta \stackrel{\text{def}}{=} \frac{1}{k_B T}.$$

The denominator $Z(\beta)$, is the partition function of a single mode and makes P_r properly normalized:

$$Z(\beta) = \sum_{r=0}^{\infty} e^{-\beta E(r)} = \frac{e^{-\beta\varepsilon/2}}{1 - e^{-\beta\varepsilon}}.$$

Here we have implicitly defined,

$$\varepsilon \stackrel{\text{def}}{=} \frac{hc}{2L}\sqrt{n_1^2 + n_2^2 + n_3^2},$$

which is the energy of a single photon. As explained here, the average energy in a mode can be expressed in terms of the partition function:

$$\langle E \rangle = -\frac{d\log(Z)}{d\beta} = \frac{\varepsilon}{2} + \frac{\varepsilon}{e^{\beta\varepsilon} - 1}.$$

This formula, apart from the first vacuum energy term, is a special case of the general formula for particles obeying Bose–Einstein statistics. Since there is no restriction on the total number of photons, the chemical potential is zero.

If we measure the energy relative to the ground state, the total energy in the box follows by summing $\langle E \rangle - \frac{\varepsilon}{2}$ over all allowed single photon states. This can be done exactly in the thermodynamic limit as L approaches infinity. In this limit, ε becomes continuous and we can then integrate $\langle E \rangle - \frac{\varepsilon}{2}$ over this parameter. To calculate the energy in the box in this way, we need to evaluate how many photon states there are in a given energy range. If we write the total number of single photon states with energies between ε and

ε + dε as $g(\varepsilon)$dε, where $g(\varepsilon)$ is the density of states (which is evaluated below), then we can write:

$$U=\int_0^\infty \frac{\varepsilon}{e^{\beta\varepsilon}-1}g(\varepsilon)d\varepsilon.$$

To calculate the density of states we rewrite equation $E_{n_1,n_2,n_3}(r)=\left(r+\frac{1}{2}\right)\frac{hc}{2L}\sqrt{n_1^2+n_2^2+n_3^2}$ as follows:

$$\varepsilon \overset{\text{def}}{=} \frac{hc}{2L}n,$$

where n is the norm of the vector n = (n_1, n_2, n_3):

$$n=\sqrt{n_1^2+n_2^2+n_3^2}.$$

For every vector n with integer components larger than or equal to zero, there are two photon states. This means that the number of photon states in a certain region of n-space is twice the volume of that region. An energy range of dε corresponds to shell of thickness $dn=\frac{2L}{hc}$ dε in n-space. Because the components of n have to be positive, this shell spans an octant of a sphere. The number of photon states $g(\varepsilon)$dε, in an energy range dε, is thus given by:

$$g(\varepsilon)d\varepsilon=2\frac{1}{8}4\pi n^2\,dn=\frac{8\pi L^3}{h^3c^3}\varepsilon^2\,d\varepsilon.$$

Inserting this in $U=\int_0^\infty \frac{\varepsilon}{e^{\beta\varepsilon}-1}g(\varepsilon)d\varepsilon$ gives:

$$U=L^3\frac{8\pi}{h^3c^3}\int_0^\infty \frac{\varepsilon^3}{e^{\beta\varepsilon}-1}d\varepsilon.$$

From this equation one can derive the spectral energy density as a function of frequency $u_\nu(T)$ and as a function of wavelength $u_\lambda(T)$:

$$\frac{U}{L^3}=\int_0^\infty u_\nu(T)d\nu,$$

where,

$$u_\nu(T)=\frac{8\pi h\nu^3}{c^3}\frac{1}{e^{h\nu/k_BT}-1}.$$

and,

$$\frac{U}{L^3}=\int_0^\infty u_\lambda(T)d\lambda,$$

where

$$u_\lambda(T) = \frac{8\pi hc}{\lambda^5} \frac{1}{e^{hc/\lambda k_B T} - 1}.$$

This is also a spectral energy density function with units of energy per unit wavelength per unit volume. Integrals of this type for Bose and Fermi gases can be expressed in terms of polylogarithms. In this case, however, it is possible to calculate the integral in closed form using only elementary functions. Substituting

$$\varepsilon = k_B T x,$$

in $U = L^3 \frac{8\pi}{h^3 c^3} \int_0^\infty \frac{\varepsilon^3}{e^{\beta\varepsilon} - 1} d\varepsilon$, makes the integration variable dimensionless giving:

$$u(T) = \frac{8\pi (k_B T)^4}{(hc)^3} J,$$

where J is a Bose–Einstein integral given by:

$$J = \int_0^\infty \frac{x^3}{e^x - 1} dx = \frac{\pi^4}{15}.$$

The total electromagnetic energy inside the box is thus given by:

$$\frac{U}{V} = \frac{8\pi^5 (k_B T)^4}{15(hc)^3},$$

where $V = L^3$ is the volume of the box.

The combination $\frac{hc}{k_B}$ has the value 14387.770 μm·K.

This is not the Stefan–Boltzmann law (which provides the total energy *radiated* by a black body per unit surface area per unit time), but it can be written more compactly using the Stefan–Boltzmann constant σ, giving

$$\frac{U}{V} = \frac{4\sigma T^4}{c}.$$

The constant $\frac{4\sigma}{c}$ is sometimes called the radiation constant.

Since the radiation is the same in all directions, and propagates at the speed of light (c), the spectral radiance of radiation exiting the small hole is

$$B_\nu(T) = \frac{u_\nu(T) c}{4\pi},$$

which yields,

$$B_\nu(T) = \frac{2h\nu^3}{c^2} \frac{1}{e^{h\nu/k_B T} - 1}.$$

It can be converted to an expression for $B_\lambda(T)$ in wavelength units by substituting ν by c/λ and evaluating,

$$B_\lambda(T) = B_\nu(T) \left| \frac{d\nu}{d\lambda} \right|.$$

Dimensional analysis shows that the unit of steradians, shown in the denominator of the right hand side of the equation above, is generated in and carried through the derivation but does not appear in any of the dimensions for any element on the left-hand-side of the equation.

Wein's Displacement Law

Wien's displacement law states that the black-body radiation curve for different temperatures will peak at different wavelengths that are inversely proportional to the temperature. The shift of that peak is a direct consequence of the Planck radiation law, which describes the spectral brightness of black-body radiation as a function of wavelength at any given temperature. However, it had been discovered by Wilhelm Wien several years before Max Planck developed that more general equation, and describes the entire shift of the spectrum of black-body radiation toward shorter wavelengths as temperature increases.

Formally, Wien's displacement law states that the spectral radiance of black-body radiation per unit wavelength, peaks at the wavelength λ_{max} given by:

$$\lambda_{max} = \frac{b}{T}$$

where T is the absolute temperature in kelvins. b is a constant of proportionality called *Wien's displacement constant*, equal to 2.897771955...×10⁻³ m·K, or to obtain wavelength in micrometers, $b \approx 2898$ μm·K. If one is considering the peak of black body emission per unit frequency or per proportional bandwidth, one must use a different proportionality constant. However, the form of the law remains the same: the peak wavelength is inversely proportional to temperature, and the peak frequency is directly proportional to temperature.

Wien's displacement law may be referred to as "Wien's law", a term which is also used for the Wien approximation.

Examples:

Wien's displacement law is relevant to some everyday experiences:

- A piece of metal heated by a blow torch first becomes "red hot" as the very longest visible wavelengths appear red, then becomes more orange-red as the temperature is increased, and at very high temperatures would be described as "white hot" as shorter and shorter wavelengths come to predominate the black body emission spectrum. Before it had even reached the red hot temperature, the thermal emission was mainly at longer infrared wavelengths, which are not visible; nevertheless, that radiation could be felt as it warms one's nearby skin.
- One easily observes changes in the color of an incandescent light bulb (which produces light through thermal radiation) as the temperature of its filament is varied by a light dimmer. As the light is dimmed and the filament temperature decreases, the distribution of color shifts toward longer wavelengths and the light appears redder, as well as dimmer.
- A wood fire at 1500 K puts out peak radiation at about 2000 nm. 98% of its radiation is at wavelengths longer than 1000 nm, and only a tiny proportion at visible wavelengths (390–700 nm). Consequently, a campfire can keep one warm but is a poor source of visible light.
- The effective temperature of the Sun is 5778 K. Using Wien's law, one finds a peak emission per nanometer (of wavelength) at a wavelength of about 500 nm, in the green portion of the spectrum near the peak sensitivity of the human eye. On the other hand, in terms of power per unit optical frequency, the Sun's peak emission is at 343 THz or a wavelength of 883 nm in the near infrared. In terms of power per percentage bandwidth, the peak is at about 635 nm, a red wavelength. Regardless of how one wants to plot the spectrum, about half of the sun's radiation is at wavelengths shorter than 710 nm, about the limit of the human vision. Of that, about 12% is at wavelengths shorter than 400 nm, ultraviolet wavelengths, which is invisible to an unaided human eye. It can be appreciated that a rather large amount of the Sun's radiation falls in the fairly small visible spectrum.

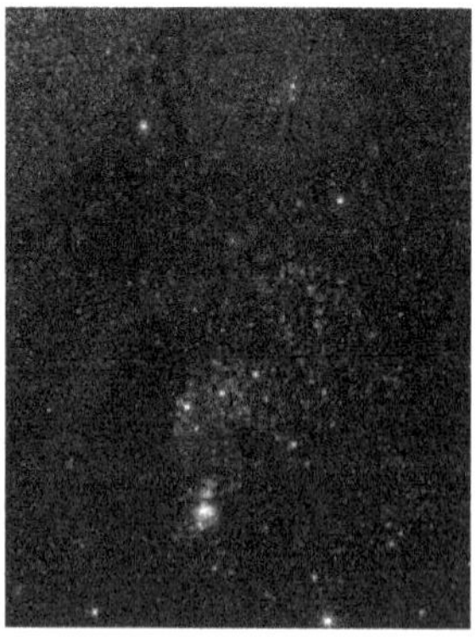

The color of a star is determined by its temperature, according to Wien's law.

In the constellation of Orion, one can compare Betelgeuse ($T \approx 3300$ K, upper left), Rigel ($T = 12100$ K, bottom right), Bellatrix ($T = 22000$ K, upper right), and Mintaka ($T = 31800$ K, rightmost of the 3 "belt stars" in the middle).

- The preponderance of emission in the visible range, however, is not the case in most stars. The hot supergiant Rigel emits 60% of its light in the ultraviolet, while the cool supergiant Betelgeuse emits 85% of its light at infrared wavelengths. With both stars prominent in the constellation of Orion, one can easily appreciate the color difference between the blue-white Rigel ($T = 12100$ K) and the red Betelgeuse ($T \approx 3300$ K). While few stars are as hot as Rigel, stars cooler than the sun or even as cool as Betelgeuse are very commonplace.

- Mammals with a skin temperature of about 300 K emit peak radiation at around 10 μm in the far infrared. This is therefore the range of infrared wavelengths that pit viper snakes and passive IR cameras must sense.

- When comparing the apparent color of lighting sources (including fluorescent lights, LED lighting, computer monitors, and photoflash), it is customary to cite the color temperature. Although the spectra of such lights are not accurately described by the black-body radiation curve, a color temperature is quoted for which black-body radiation would most closely match the subjective color of that source. For instance, the blue-white fluorescent light sometimes used in an office may have a color temperature of 6500 K, whereas the reddish tint of a dimmed incandescent light may have a color temperature (and an actual filament temperature) of 2000 K. Note that the informal description of the former (bluish) color as "cool" and the latter (reddish) as "warm" is exactly opposite the actual temperature change involved in black-body radiation.

Frequency-dependent Formulation

For spectral flux considered per unit frequency $d\upsilon$ (in hertz), Wien's displacement law describes a peak emission at the optical frequency V_{max} given by:

$$\nu_{\max} = \frac{\alpha}{h} kT \approx (5.879 \times 10^{10} \text{ Hz/K}) \cdot T$$

where $\alpha \approx 2.8214391...$ is a constant resulting from the numerical solution of the maximization equation, k is the Boltzmann constant, h is the Planck constant, and T is the temperature (in kelvins). With the emission now considered per unit frequency, this peak now corresponds to a wavelength 70% longer than the peak considered per unit wavelength.

Derivation from Planck's Law

Planck's law for the spectrum of black-body radiation predicts the Wien displacement law and may be used to numerically evaluate the constant relating temperature and peak wavelength (or frequency). According to one form of that law, the black body spectral radiance (power per emitting area per solid angle).

$$u_{\lambda}(\lambda,T)=\frac{2hc^2}{\lambda^5}\frac{1}{e^{hc/\lambda kT}-1}.$$

Differentiating $u(\lambda,T)$ with respect to λ and setting the derivative equal to zero gives,

$$\frac{\partial u}{\partial \lambda}=2hc^2\left(\frac{hc}{kT\lambda^7}\frac{e^{hc/\lambda kT}}{\left(e^{hc/\lambda kT}-1\right)^2}-\frac{1}{\lambda^6}\frac{5}{e^{hc/\lambda kT}-1}\right)=0,$$

which can be simplified to give,

$$\frac{hc}{\lambda kT}\frac{e^{hc/\lambda kT}}{e^{hc/\lambda kT}-1}-5=0.$$

By defining:

$$x\equiv\frac{hc}{\lambda kT},$$

the equation becomes one in the single variable x:

$$\frac{xe^x}{e^x-1}-5=0.$$

which is equivalent to:

$$(x-5)e^x+5=0.$$

This equation is easily numerically solved using Newton's method yielding x = 4.965114231744276 to double precision floating point accuracy.

Solving for the wavelength λ in units of nanometers, and using kelvin for the temperature yields:

$$\lambda_{\max}=\frac{hc}{x}\frac{1}{kT}=\frac{2.897771955185172\times10^6\ \mathrm{nm\cdot K}}{T}.$$

Alternate Maxima

The form of Wien's displacement law in terms of maximum radiance per unit *frequency* is derived using similar methods, but starting with the form of Planck's law as a function of frequency ν:

$$u_\nu(\nu,T) = \frac{2h\nu^3}{c^2}\frac{1}{e^{h\nu/kT}-1}.$$

The preceding process using this equation yields:

$$-\frac{h\nu}{kT}\frac{e^{h\nu/kT}}{e^{h\nu/kT}-1}+3=0.$$

The net result is:

$$(x-3)e^x+3=0.$$

This is similarly solved with Newton's method yielding x = 2.8214393721220787 to double precision floating point accuracy. Solving for ν produces:

$$\nu_{\max} = \frac{xk}{h}T = (0.0587892381136085\,5\,\mathrm{THz}\cdot\mathrm{K}^{-1})\cdot T.$$

Notice that for a given temperature this form implies a different maximal wavelength, for these functions are power *density* functions, and take different forms for different parameterizations. Since wavelength and frequency have a reciprocal relation, they constitute significantly different parameterizations.

For example, using T = 6000 K the wavelength for maximal power density is λ = 482.962 nm with corresponding frequency ν = 620.737 THz. For the same temperature, the frequency for maximal power density is ν = 352.735 THz with corresponding wavelength λ = 849.907 nm.

The total radiated power is the integral of the power distribution over all positive values (for these two cases), and that is invariant for a given temperature. Additionally, for a given temperature the radiated power consisting of all photons between two wavelengths must be the same regardless of which distribution you use; that is to say, integrating the wavelength distribution from λ_1 to λ_2 will result in the same value as integrating the frequency distribution between the two frequencies that correspond to λ_1 and λ_2, namely from c/λ_2 to c/λ_1. However, the distribution *shape* depends on the parameterization, and for a different parameterization the distribution will typically have a different peak power density, as these calculations demonstrate.

Using the value 4 to solve the implicit equation yields the peak in the spectral radiance density function expressed in the parameter radiance *per proportional bandwidth*. This is perhaps a more intuitive way of presenting "wavelength of peak emission". That yields x = 3.9206903948728864 to double precision floating point accuracy.

The important point of Wien's law, however, is that *any* such wavelength marker, including the median wavelength (or the wavelength below which a specified percentage of the emission occurs) is proportional to the reciprocal of temperature. That is, the shape of the distribution for a given parameterization, scales with and translates according to temperature, and can be calculated once for a canonical temperature, then appropriately shifted and scaled to obtain the distribution for another temperature. This is a consequence of the strong statement of Wien's law.

Kirchhoff's Law of Thermal Radiation

In general, both the emissivity, ε, and the absorptivity, α, of a surface depend on the temperature and the wavelength of the radiation. Kirchhoff's law of thermal radiation, postulated by a German physicist Gustav Robert Kirchhoff, states that the emissivity and the absorptivity of a surface at a given temperature and wavelength are equal.

For an arbitrary body emitting and absorbing thermal radiation in thermodynamic equilibrium, the emissivity is equal to the absorptivity.

$$\text{emissivity } \varepsilon = \text{absorptivity } \alpha$$

This law must be also valid in order to satisfy the Second Law of Thermodynamics. As was written, all bodies above absolute zero temperature radiate some heat. Two objects radiate heat toward each other. But what if a colder object with high emissivity radiates toward a hotter object with very low emissivity? This seems to violate the Second Law of Thermodynamics, which states that heat cannot spontaneously flow from cold system to hot system without external work being performed on the system. The paradox is resolved by the fact that each body must be in direct line of sight of the other to receive radiation from it. Therefore, whenever the cool body is radiating heat to the hot body, the hot body must also be radiating heat to the cool body. Moreover, the hot body will radiate more energy than cold body. The case of different emissivities is solved by the Kirchhoff's Law of thermal radiation, which states that object with low emissivity have also low absorptivity. As a result, heat cannot spontaneously flow from cold system to hot system and the second law is still satisfied.

The emissivity, ε, of the surface of a material is its effectiveness in emitting energy as thermal radiation and varies between 0.0 and 1.0. emissivity of various material. By definition, a blackbody in thermal equilibrium has an emissivity of ε = 1.0. Real objects do not radiate as much heat as a perfect black body. They radiate less heat than a black body and therefore are called gray bodies. To take into account the fact that real objects are gray bodies, the Stefan-Boltzmann law must include emissivity. Quantitatively, emissivity is the ratio of the thermal radiation from a surface to the radiation from an ideal black surface at the same temperature as given by the Stefan–Boltzmann law. Emissivity is simply a factor by which we multiply the black body heat transfer to take into account that the black body is the ideal case.

The surface of a blackbody emits thermal radiation at the rate of approximately 448 watts per square metre at room temperature (25 °C, 298.15 K). Real objects with emissivities less than 1.0 (e.g. copper wire) emit radiation at correspondingly lower rates (e.g. 448 x 0.03 = 13.4 W/m2). Emissivity plays important role in heat transfer problems. For example, solar heat collectors incorporate selective surfaces that have very low emissivities. These collectors waste very little of the solar energy through emission of thermal radiation.

Another important radiation property of a surface is its absorptivity, α, which is the fraction of the radiation energy incident on a surface that is absorbed by the surface. Like emissivity, value of absorptivity is in the range $0 < \alpha < 1$. From its definition, a blackbody, which is an idealized physical body, absorbs all incident electromagnetic radiation, regardless of frequency or angle of incidence. That is, a blackbody is a perfect absorber. Since for real objects the absorptivity is less than unity, a real object can not absorb all incident light. The incomplete absorption can be due to some of the incident light being transmitted through the body or to some of it being reflected at the surface of the body.

In general, the absorptivity and the emissivity are interconnected by the Kirchhoff's Law of thermal radiation, which states:

For an arbitrary body emitting and absorbing thermal radiation in thermodynamic equilibrium, the emissivity is equal to the absorptivity.

emissivity ε = absorptivity α

Note that visible radiation occupies a very narrow band of the spectrum from 0.4 to 0.76 nm, we cannot make any judgments about the blackness of a surface on the basis of visual observations. For example, consider white paper that reflects visible light and thus appear white. On the other hand it is essentially black for infrared radiation (absorptivity $\alpha = 0.94$) since they strongly absorb long-wavelength radiation.

Stefan–Boltzmann Law

The Stefan–Boltzmann law describes the power radiated from a black body in terms of its temperature. Specifically, the Stefan–Boltzmann law states that the total energy radiated per unit surface area of a black body across all wavelengths per unit time $j^{\star}$ (also known as the black-body *radiant emittance*) is directly proportional to the fourth power of the black body's thermodynamic temperature T:

$$j^{\star} = \sigma T^{4}.$$

The constant of proportionality σ, called the Stefan–Boltzmann constant, is derived from other known physical constants. The value of the constant is,

$$\sigma = \frac{2\pi^5 k^4}{15c^2 h^3} = 5.670373 \times 10^{-8}\,\mathrm{W\,m^{-2}K^{-4}},$$

where k is the Boltzmann constant, h is Planck's constant, and c is the speed of light in a vacuum. The radiance (watts per square metre per steradian) is given by,

$$L = \frac{j^\star}{\pi} = \frac{\sigma}{\pi} T^4.$$

A body that does not absorb all incident radiation (sometimes known as a grey body) emits less total energy than a black body and is characterized by an emissivity, $\varepsilon < 1$:

$$j^\star = \varepsilon \sigma T^4.$$

The radiant emittance $j^\star$ has dimensions of energy flux (energy per time per area), and the SI units of measure are joules per second per square metre, or equivalently, watts per square metre. The SI unit for absolute temperature T is the kelvin. ε is the emissivity of the grey body; if it is a perfect blackbody, $\varepsilon = 1$. In the still more general (and realistic) case, the emissivity depends on the wavelength, $\varepsilon = \varepsilon(\lambda)$.

To find the total power radiated from an object, multiply by its surface area,

$$P = Aj^\star = A\varepsilon\sigma T^4.$$

Wavelength- and subwavelength-scale particles, metamaterials, and other nanostructures are not subject to ray-optical limits and may be designed to exceed the Stefan–Boltzmann law.

Temperature of the Sun

With his law Stefan also determined the temperature of the Sun's surface. He inferred from the data of Jacques-Louis Soret (1827–1890) that the energy flux density from the Sun is 29 times greater than the energy flux density of a certain warmed metal lamella (a thin plate). A round lamella was placed at such a distance from the measuring device that it would be seen at the same angle as the Sun. Soret estimated the temperature of the lamella to be approximately 1900 °C to 2000 °C. Stefan surmised that ⅓ of the energy flux from the Sun is absorbed by the Earth's atmosphere, so he took for the correct Sun's energy flux a value 3/2 times greater than Soret's value, namely 29 × 3/2 = 43.5.

Precise measurements of atmospheric absorption were not made until 1888 and 1904. The temperature Stefan obtained was a median value of previous ones, 1950 °C and the absolute thermodynamic one 2200 K. As $2.57^4 = 43.5$, it follows from the law that the temperature of the Sun is 2.57 times greater than the temperature of the lamella, so Stefan got a value of 5430 °C or 5700 K (the modern value is 5778 K). This was the first sensible value for the temperature of the Sun. Before this, values ranging from as low as 1800 °C to as high as 13,000,000 °C were claimed. The lower value of 1800 °C was determined by Claude Pouillet (1790–1868) in 1838 using the Dulong–Petit law. Pouillet also took just half the value of the Sun's correct energy flux.

Temperature of Stars

The temperature of stars other than the Sun can be approximated using a similar means by treating the emitted energy as a black body radiation. So:

$$L = 4\pi R^2 \sigma T_e^4$$

where L is the luminosity, σ is the Stefan–Boltzmann constant, R is the stellar radius and T is the effective temperature. This same formula can be used to compute the approximate radius of a main sequence star relative to the sun:

$$\frac{R}{R_\odot} \approx \left(\frac{T_\odot}{T}\right)^2 \cdot \sqrt{\frac{L}{L_\odot}}$$

where $R_\odot$ is the solar radius, $L_\odot$ is the solar luminosity, and so forth.

With the Stefan–Boltzmann law, astronomers can easily infer the radii of stars. The law is also met in the thermodynamics of black holes in so-called Hawking radiation.

Effective Temperature of the Earth

Similarly we can calculate the effective temperature of the Earth $T_\oplus$ by equating the energy received from the Sun and the energy radiated by the Earth, under the black-body approximation (Earth's own production of energy being small enough to be negligible). The luminosity of the Sun, $L_\odot$, is given by:

$$L_\odot = 4\pi R_\odot^2 \sigma T_\odot^4$$

At Earth, this energy is passing through a sphere with a radius of a_0, the distance between the Earth and the Sun, and the irradiance (received power per unit area) is given by,

$$E_\oplus = \frac{L_\odot}{4\pi a_0^2}$$

The Earth has a radius of $R_\oplus$, and therefore has a cross-section of $\pi R_\oplus^2$. The radiant flux (i.e. solar power) absorbed by the Earth is thus given by:

$$\Phi_{\text{abs}} = \pi R_\oplus^2 \times E_\oplus$$

Because the Stefan–Boltzmann law uses a fourth power, it has a stabilizing effect on the exchange and the flux emitted by Earth tends to be equal to the flux absorbed, close to the steady state where,

$$4\pi R_{\oplus}^2 \sigma T_{\oplus}^4 = \pi R_{\oplus}^2 \times E_{\oplus}$$

$$= \pi R_{\oplus}^2 \times \frac{4\pi R_{\odot}^2 \sigma T_{\odot}^4}{4\pi a_0^2}$$

$T_{\oplus}$ can then be found:

$$T_{\oplus}^4 = \frac{R_{\odot}^2 T_{\odot}^4}{4a_0^2}$$

$$T_{\oplus} = T_{\odot} \times \sqrt{\frac{R_{\odot}}{2a_0}}$$

$$= 5780\,\mathrm{K} \times \sqrt{\frac{696 \times 10^6\ \mathrm{m}}{2 \times 149.598 \times 10^9\ \mathrm{m}}}$$

$$\approx 279\,\mathrm{K}$$

where, $T_{\odot}$ is the temperature of the Sun, $R_{\odot}$ the radius of the Sun, and a_0 is the distance between the Earth and the Sun. This gives an effective temperature of 6 °C on the surface of the Earth, assuming that it perfectly absorbs all emission falling on it and has no atmosphere.

The Earth has an albedo of 0.3, meaning that 30% of the solar radiation that hits the planet gets scattered back into space without absorption. The effect of albedo on temperature can be approximated by assuming that the energy absorbed is multiplied by 0.7, but that the planet still radiates as a black body (the latter by definition of effective temperature, which is what we are calculating). This approximation reduces the temperature by a factor of $0.7^{1/4}$, giving 255 K (−18 °C).

The above temperature is Earth's as seen from space, not ground temperature but an average over all emitting bodies of Earth from surface to high altitude. Because of the greenhouse effect, the Earth's actual average surface temperature is about 288 K (15 °C), which is higher than the 255 K effective temperature, and even higher than the 279 K temperature that a black body would have.

In the above discussion, we have assumed that the whole surface of the earth is at one temperature. Another interesting question is to ask what the temperature of a blackbody surface on the earth would be assuming that it reaches equilibrium with the sunlight falling on it. This of course depends on the angle of the sun on the surface and on how much air the sunlight has gone through. When the sun is at the zenith and the surface is horizontal, the irradiance can be as high as 1120 W/m². The Stefan–Boltzmann law then gives a temperature of,

$$T = \left(\frac{1120 \text{ W/m}^2}{\sigma} \right)^{1/4} \approx 375 \text{ K}$$

or 102 °C. (Above the atmosphere, the result is even higher: 394 K.) We can think of the earth's surface as "trying" to reach equilibrium temperature during the day, but being cooled by the atmosphere, and "trying" to reach equilibrium with starlight and possibly moonlight at night, but being warmed by the atmosphere.

Thermodynamic Derivation of the Energy Density

The fact that the energy density of the box containing radiation is proportional to T^4 can be derived using thermodynamics. This derivation uses the relation between the radiation pressure p and the internal energy density u, a relation that can be shown using the form of the electromagnetic stress–energy tensor. This relation is:

$$p = \frac{u}{3}.$$

Now, from the fundamental thermodynamic relation,

$$dU = T\,dS - p\,dV,$$

we obtain the following expression, after dividing by dV and fixing T:

$$\left(\frac{\partial U}{\partial V} \right)_T = T \left(\frac{\partial S}{\partial V} \right)_T - p = T \left(\frac{\partial p}{\partial T} \right)_V - p.$$

The last equality comes from the following Maxwell relation:

$$\left(\frac{\partial S}{\partial V} \right)_T = \left(\frac{\partial p}{\partial T} \right)_V.$$

From the definition of energy density it follows that,

$$U = uV$$

where the energy density of radiation only depends on the temperature, therefore

$$\left(\frac{\partial U}{\partial V} \right)_T = u \left(\frac{\partial V}{\partial V} \right)_T = u.$$

Now, the equality

$$\left(\frac{\partial U}{\partial V}\right)_T = T\left(\frac{\partial p}{\partial T}\right)_V - p,$$

after substitution of $\left(\frac{\partial U}{\partial V}\right)_T$ and p for the corresponding expressions, can be written as,

$$u = \frac{T}{3}\left(\frac{\partial u}{\partial T}\right)_V - \frac{u}{3}.$$

Since the partial derivative $\left(\frac{\partial u}{\partial T}\right)_V$ can be expressed as a relationship between only u and T (if one isolates it on one side of the equality), the partial derivative can be replaced by the ordinary derivative. After separating the differentials the equality becomes,

$$\frac{du}{4u} = \frac{dT}{T},$$

which leads immediately to $u = AT^4$, with A as some constant of integration.

Derivation from Planck's Law

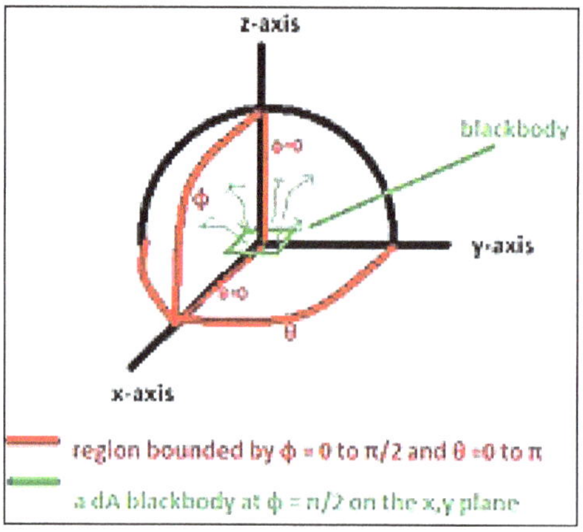

Deriving the Stefan–Boltzmann Law using the Planck's law.

The law can be derived by considering a small flat black body surface radiating out into a half-sphere. This derivation uses spherical coordinates, with θ as the zenith angle and φ as the azimuthal angle; and the small flat blackbody surface lies on the xy-plane, where $\theta = {}^{\pi}/_{2}$.

The intensity of the light emitted from the blackbody surface is given by Planck's law:

$$I(\nu,T) = \frac{2h\nu^3}{c^2}\frac{1}{e^{h\nu/(kT)} - 1}.$$

where,

- $I(\ ,T)$ is the amount of power per unit surface area per unit solid angle per unit frequency emitted at a frequency υ by a black body at temperature T.
- h is Planck's constant.
- c is the speed of light.
- k is Boltzmann's constant.

The quantity $I(v,T)\,A\,dv\,d\Omega$ is the power radiated by a surface of area A through a solid angle $d\Omega$ in the frequency range between v and $v + dv$.

The Stefan–Boltzmann law gives the power emitted per unit area of the emitting body,

$$\frac{P}{A} = \int_0^{\infty} I(v,T)dv \int \cos\theta\, d\Omega$$

Note that the cosine appears because black bodies are *Lambertian* (i.e. they obey Lambert's cosine law), meaning that the intensity observed along the sphere will be the actual intensity times the cosine of the zenith angle. To derive the Stefan–Boltzmann law, we must integrate $d\Omega = \sin(\theta)\, d\theta\, d\varphi$ over the half-sphere and integrate v from 0 to ∞.

$$\frac{P}{A} = \int_0^{\infty} I(v,T)dv \int_0^{2\pi} d\varphi \int_0^{\pi/2} \cos\theta \sin\theta\, d\theta$$

$$= \pi \int_0^{\infty} I(v,T)dv$$

Then we plug in for I:

$$\frac{P}{A} = \frac{2\pi h}{c^2} \int_0^{\infty} \frac{v^3}{e^{\frac{hv}{kT}} - 1} dv$$

To evaluate this integral, do a substitution,

$$u = \frac{hv}{kT}$$

$$du = \frac{h}{kT} dv$$

which gives:

$$\frac{P}{A} = \frac{2\pi h}{c^2} \left(\frac{kT}{h}\right)^4 \int_0^{\infty} \frac{u^3}{e^u - 1} du.$$

The integral on the right is standard and goes by many names: it is a particular case of a Bose–Einstein integral, the polylogarithm, or the Riemann zeta function $\zeta(s)$. The value of the integral is $6\zeta(4)=\frac{\pi^4}{15}$, giving the result that, for a perfect blackbody surface:

$$j^{\star}=\sigma T^4\ ,\ \sigma=\frac{2\pi^5 k^4}{15c^2h^3}=\frac{\pi^2 k^4}{60\hbar^3 c^2}.$$

Finally, this proof started out only considering a small flat surface. However, any differentiable surface can be approximated by a collection of small flat surfaces. So long as the geometry of the surface does not cause the blackbody to reabsorb its own radiation, the total energy radiated is just the sum of the energies radiated by each surface; and the total surface area is just the sum of the areas of each surface—so this law holds for all convex blackbodies, too, so long as the surface has the same temperature throughout. The law extends to radiation from non-convex bodies by using the fact that the convex hull of a black body radiates as though it were itself a black body.

Energy Density

The total energy density U can be similarly calculated, except the integration is over the whole sphere and there is no cosine, and the energy flux should be divided by the velocity c:

$$U=\frac{1}{c}\int_0^{\infty} I(\nu,T)d\nu\int d\Omega$$

Thus $\int_0^{\pi/2}\cos\theta\sin\theta d\theta$ is replaced by $\int_0^{\pi}\sin\theta d\theta$, giving an extra factor of 4.

Thus, in total:

$$U=\frac{4}{c}\sigma T^4$$

Classical Gray Body Radiation Theory

Candlelight was the first man-made light source. Much later, gas lights emerged. Then, in the 19th century, the first electric incandescent lamp was invented. 140 years later, the incandescent lamp is being replaced by the LED.

Candles, gas lights, and incandescent lamps all use thermal radiation, the only heat transfer phenomenon that works without the presence of heat-conductive media. Without this special property, we wouldn't be able to live on Earth: Solar energy would pour in by this mechanism through a vacuum.

An incandescent light bulb.

One of the most interesting stories in physics is when Max Planck discovered the formula of the spectral distribution, now called the Planck distribution, for black body radiation. The *black body* is an idealized physical body that absorbs *all* incident electromagnetic radiation; equivalently, a body that emits maximized thermal radiation.

The discovery of the black body theory opened the door to quantum mechanics. With Planck's work, we now know how much power we get from a black body object at temperature T in an equilibrium state, which is called the Stefan-Boltzmann law:

$$I = n^2 \sigma T^4 \quad (W / m^2),$$

Where $\sigma \simeq 5.67 \times 10^{-8} \quad (W / m^2.K^4)$ is the Stefan-Boltzmann constant and n is the refractive index of the media.

We now consider more practical objects, called gray bodies. A gray body is an imperfect black body; i.e., a physical object that *partially* absorbs incident electromagnetic radiation. The ratio of a gray body's thermal radiation to a black body's thermal radiation at the same temperature is called the *emissivity* of the gray body.

The emissivity of a black body is unity, while that of a gray body is larger than 0 and smaller than 1. The emissivity is a function of the geometry of the radiative surface, its physical properties, and the wavelength. So, how is emissivity determined from certain structures?

For the sake of simplicity, we will consider the diffuse gray surface, which doesn't account for the spectral dependency of the radiation. The following theory is based on Gouffé's paper, which outlines the classical gray body radiation theory. The math is something that you might have learned in middle or high school, but the physical concept is a little more complicated. Gouffé's paper was cited by many researchers at the time it was published, we will try to include what is missing from the original paper in order to provide a complete explanation.

First of all, let's define our terms and conventions. When we mention reflectivity and absorptivity, we must differentiate between the "material" and "apparent" quantities. This is very important in order to avoid confusion when learning about thermal radiation. Because gray bodies typically have certain (tiny) structures on their surfaces, the "apparent" quantity viewed from a far field is different from the raw material quantity. For example, the apparent reflectivity of a rough surface is always lower than its material reflectivity.

Our goal is to calculate the apparent emissivity from a given material reflectivity and the geometry of the structure. Here, we will differentiate "material" quantities from "apparent" quantities by appending "o" in the notation. Otherwise, it is understood that we are talking about a quantity in general. The nomenclature of the terms is as follows:

- Material reflectivity, ρ_0.
- Material absorptivity, α_0.
- Material emissivity, ε_0.
- Apparent reflectivity, ρ.
- Apparent absorptivity, α.
- Apparent emissivity, ε.
- Cavity's internal area, S.
- Opening area, A.
- View angle, θ.
- View factor (normalized solid angle), F.
- View factor (area ratio), G.

From the conservation law of energy, we have the following relation for opaque materials between the reflectivity ρ and the absorptivity α:

$$1=\rho+\alpha$$

Kirchhoff's law states that the material emissivity ε at thermodynamic equilibrium is equal to the absorptivity α; i.e.,

$$\varepsilon=\alpha$$

From the two relations above, we get the emissivity from the reflectivity; i.e.,

$$\varepsilon=1-\rho$$

We consider a single surface structure, as shown in the following figure. The structure can be anything, here we use a spherical cavity with an opening on top with area A (circle of radius R) at distance L from the bottom of the cavity. We assume that the material reflectivity ρ_0 is uniform over the internal surface of the cavity and the reflection takes place according to Lambert's law; i.e., the intensity of the reflection is $Ir = \rho_0 \cos\theta$, where θ is the viewing angle, as depicted in the figure. We want to calculate how much apparent reflection we get out of an incident light of the energy of unity.

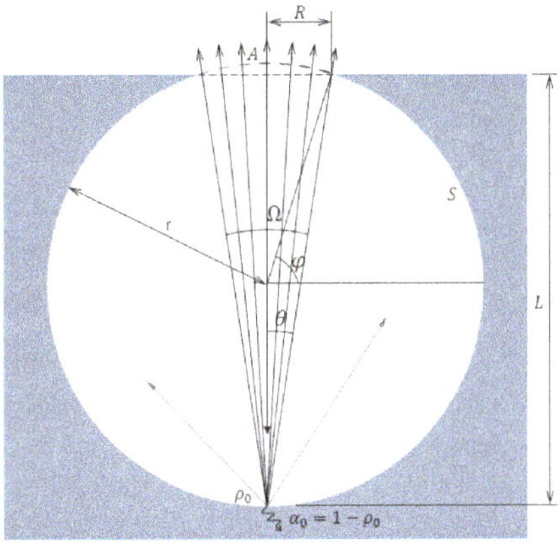

Surface structure for calculating the apparent reflectivity.

As the first-order approximation, the reflection from the bottom of the cavity is:

$$\rho_o F$$

where F is called the view factor.

Note that in Gouffé's paper, the view factor is assumed to be uniform over the cavity surface. In the COMSOL software, it is instead a function of the position, which is always true. (So, we need integration to compute the emissivity.)

In this case, the view factor is the normalized solid angle to the opening from the first reflection point. The solid angle Ω is calculated as:

$$\Omega = \int_0^{\theta}\int_0^{2\pi} \sin\theta' \cos\theta' \, d\theta' \, d\varphi' = \pi \sin^2\theta$$

Note that the total solid angle for the hemisphere is π, not 2π. This is due to the Lambertian factor, $\cos\theta$.

As a result, the view factor F is,

$$F = \sin^2\theta$$

To this approximation, the apparent reflectivity is,

$$\rho^{(1)} = \rho_0 F$$

from which the apparent emissivity is derived as,

$$\varepsilon^{(1)} = 1 - \rho_0 F$$

Roughly speaking, the smaller the opening, the more the cavity becomes a black body due to the view factor.

Next, let's improve the approximation. After the first reflection, which we already calculated, the rest is absorbed by the cavity material or contributes to further reflections. The material absorption α_0 is $\alpha_0 = 1 - \rho_0$ from the conservation of energy. The energy left for the subsequent reflections is:

$$1 - \rho_0 F - \alpha_0 = \rho_0 (1 - F)$$

Now, assuming again that further reflections take place in a *uniform* way, the reflection that gets out of the cavity at the second reflection should be the above quantity multiplied by the material reflectivity one more time and another view factor defined by the area ratio, $G = A / S$, where S is the area of the cavity, including the opening area A. The apparent reflectivity for the second reflection is:

$$\rho_0^2 (1 - F) G$$

Similarly to the first-order approximation, the apparent emissivity for the second-order approximation is:

$$\varepsilon^{(2)} = 1 - \rho_0 F - \rho_0^2 (1 - F) G$$

To this approximation, the apparent emissivity should become more accurate with the additional term.

Finally, we can take all of the reflections into account by calculating the following converging infinite series:

$$\varepsilon^{(\infty)} = 1 - \rho_0 F - \rho_0^2 (1 - F) G - \rho_0^3 (1 - F) G (1 - G) - \rho_0^4 (1 - F) G (1 - G)^2 - \ldots$$

$$= \frac{(1 - \rho_0)(1 + \rho_0 (G - F))}{1 - \rho_0 (1 - G)}$$

In the case of a sphere, it can be shown that $F = G$, which reduces this result to,

$$\varepsilon^{(\infty)} = \frac{1 - \rho_0}{1 - \rho_0 (1 - G)}.$$

Now, let's rewrite F and G explicitly in terms of the geometry parameters R and L. From the geometry, it's easy to prove that $\theta = R / \sqrt{L^2 + R^2}$ from which F can be rewritten as,

$$F = \sin^2 \theta = \frac{R^2}{R^2 + L^2} = \frac{1}{1 + \left(\frac{L}{R}\right)^2}.$$

Next, the opening area A can be calculated by,

$$A = \int_{\varphi}^{\pi/2} 2\pi r^2 \cos\varphi' d\varphi' = 2\pi r^2 (1 - \sin\varphi)$$

where, r is the radius of the sphere and φ is the angle between the plane intersecting the sphere center, which is parallel to the opening perimeter.

From the geometry, there are relations that $R^2 + (L - r)^2 = r^2$, which yields that $r = (R^2 + L^2)/(2L)$ and $\cos\varphi = R/r$. By using these relations, the opening area A can be rewritten as,

$$A = 2\pi r(2r - L)$$

Since the total sphere area is $S = 4\pi r^2$, the view factor G is given as,

$$G = \frac{A}{S} = \frac{2\pi r(2r - L)}{4\pi r^2} = \frac{1}{1 + \left(\frac{L}{R}\right)^2},$$

which proves that $F = G$.

COMSOL Multiphysics Simulation versus Theory

So far, we've learned the classical gray body theories. Now, let's compute our final goal, the apparent emissivity for a gray body consisting of a spherical shell with an opening, by using the Heat Transfer with Surface-to-Surface Radiation interface. Then, we can compare the computed dependency of the opening size on the emissivity with the approximate formulas.

Before moving to the main results, let's check one thing. In COMSOL Multiphysics, we can calculate the view factor values (corresponding to G in the above theory, but as a function of position) using the radopu and radopd operators. The quantity that is obtained by using these operands is purely geometric, so there is no need to run a study to use this operator. If you change the geometry, you can just update the solution before using the operator.

In this model, there is only one radiating surface because the model is an open cavity. So, COMSOL Multiphysics can calculate only the view factor for the cavity surface *itself*. We'll call this the *self view factor*. The view factor G, which we have discussed so far as the factor that views outside of the cavity from a point on the cavity surface, can be called the *ambient view factor*. We can calculate the view factor G by subtracting the self view factor from unity; i.e., 1 intop1 (comp1.ht.radopu (1,0))/intop1.

The view factor is position dependent in general, and so is the view factor in the

COMSOL software. To calculate the contribution from all points on the cavity surface, we need to integrate the view factor as a function of position. The operator intop1 is an integration operator defined on the cavity surface. The following figure compares the calculated results with the classical view factor theory, showing very good agreement.

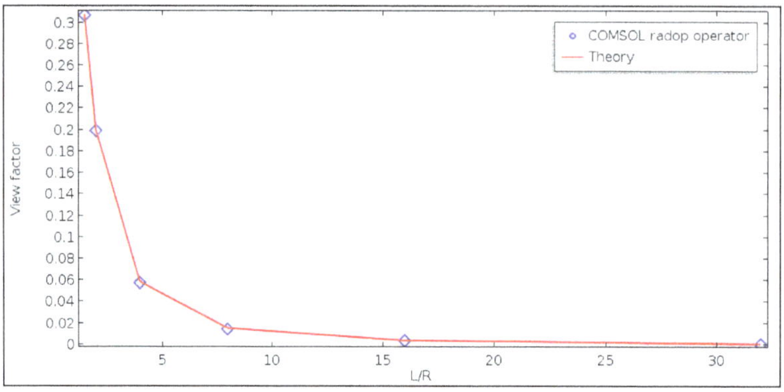

Comparison between the classical theory and the radopu operator calculation in COMSOL Multiphysics for the ambient view factor.

Now, let's move on to the settings. To calculate surface-to-surface radiation, a Surface-to-Surface Radiation physics interface and a Diffuse Surface node for the internal surface of the sphere need to be added, and a Heat Transfer with Surface-to-Surface Radiation multiphysics coupling node needs to be added. COMSOL Multiphysics will compute the view factor by performing surface integration for each mesh node except when the geometry is axisymmetric where a closed-form expression is used. For the sake of simplicity, we don't include ambient radiation; i.e., $T_{amb} = 0$. A *Temperature* boundary condition is set to the outer boundary of the spherical shell with T_O = 2500 deg K. The material emissivity is set to 0.5 in order to see the difference more easily.

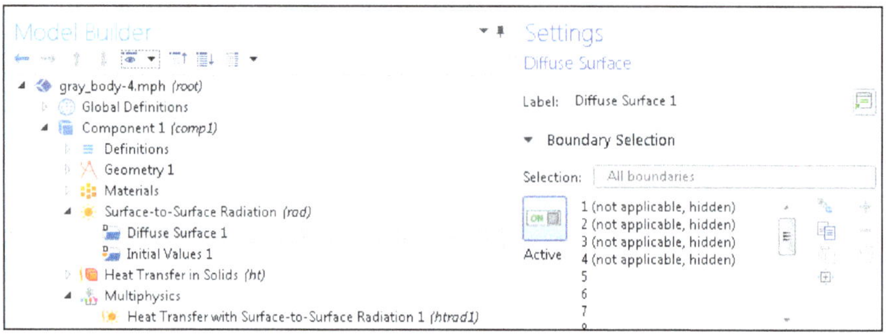

Settings for the Diffuse Surface node.

The following plots show the computed ambient view factor, radiosity, and temperature for various opening radii. Following Gouffé's paper, the geometry aspect ratio defined by L / R, where R is the opening's radius and L is the cavity height, is used as a sweep parameter.

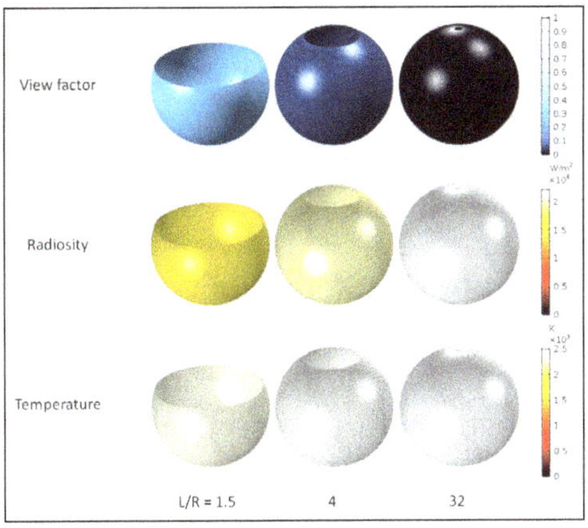

Computational results for the ambient view factor, radiosity, and temperature versus L / R.

These results qualitatively suggest that the smaller the size of the opening, the more the cavity emits thermal radiation and the higher the surface temperature.

To calculate the apparent emissivity for the computed radiosity result, we can now use the Stefan-Boltzmann law. The radiosity of the black body at temperature T_O *is* σT_0^4. Dividing the computed radiosity by this number, we get the apparent emissivity of the gray body spherical cavity, which is intop1(ht.J)/intop1(1)/(sigma_const*T0^4). Let's now compare the results with the various theories we have discussed so far.

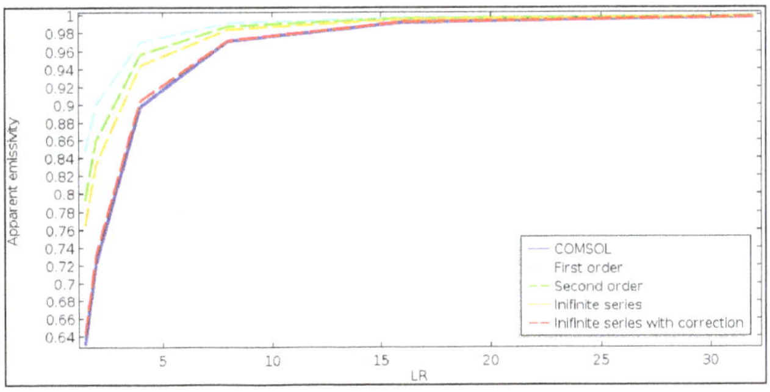

Comparison between COMSOL Multiphysics results and various theoretical approximations for the apparent emissivity for a spherical cavity.

The cyan line shows the first-order approximation, which underestimates the amount of reflection causing the higher emissivity, which is inaccurate. The second order obviously improves the accuracy, but it's not enough. The plot for the infinite series approximation (orange line) gives a much better result, but it's still not very close to the COMSOL Multiphysics results.

The reason for this discrepancy is mentioned in the last part of Gouffé's paper. As some readers might have noticed in the previous figure, the surface temperature is different for each geometry, regardless of the same outer boundary temperature of 2500 deg K.

This is called radiation cooling. We have yet to include this effect, which takes place for any real material with a finite thermal conductivity. The effect changes the temperature of the inner surface of the shell temperature, depending on the opening area. Therefore, the temperature needs to be compensated in order to calculate the emissivity at the same temperature for all opening areas.

The correction factor is, owing to the Stefan-Boltzmann law, the fourth power of the temperature ratio; i.e., (maxop1(T))^4/T0^4, where maxop1 is an operator defined on the cavity surface that finds the maximum value on the surface. Finally, with this correction, the red curve is the most accurate theoretical prediction, which agrees very well with the COMSOL Multiphysics results (blue line).

Designing an Efficient Incandescent Lamp with the COMSOL Software

The light source in an incandescent lamp is created by a twisted tungsten filament. The material emissivity of tungsten is 0.462 at 2500 deg K for a 0.467-um wavelength. In the past, researchers proposed that we can design more efficient incandescent lamps if we fabricate some microstructures on the surface of the tungsten filament. This is true. As we just learned, the apparent emissivity can be close to 1 if we have a cavity with a very tiny opening, which we can call a black body filament. In addition, research also proposed that if the maximum cavity size is about a half of 0.78 um, then any infrared light with a wavelength larger than 0.78 um may be suppressed due to the waveguide's cutoff effect. Then, the efficiency for the visible light may be significantly improved.

This effect may be more than the emissivity enhancement, because thermal radiation below tungsten's melting temperature consists of mostly infrared light that is wasted to heat. It would be fantastic if we could really cut the infrared off.

"Dream" incandescent lamp with an infrared suppressing black body filament.

Unfortunately, this "dream lamp" can't be a reality for various reasons.

First, it isn't possible for the surface to be made up of "all" holes. To make holes with a low view factor, meaning a smaller opening area than the hole size, the surface can't be densely filled with the openings; i.e., the surface needs a flatter area between the openings, which radiate infrared. Alternatively, we can make deep holes to decrease the ambient view factor and then make each hole as close to the adjacent holes as possible. However, this makes it more difficult to fabricate deeper holes.

Also, the surface energy of the surface with holes seems to be higher than that of a flat surface. So, the surface with holes tends to melt down at a lower temperature than the bulk melting temperature.

Third, it is very difficult to make holes on a twisted wire filament because it's a 3D structure. It is easier to make holes on a flat ribbon filament because it's 2D, but flat filaments are not electrically convenient because they need more current to achieve the same temperature than the twisted wire filament (higher voltage and lower current is convenient for our current infrastructure).

THERMAL RADIATION

Thermal radiation is electromagnetic radiation generated by the thermal motion of particles in matter. All matter with a temperature greater than absolute zero emits thermal radiation. Particle motion results in charge-acceleration or dipole oscillation which produces electromagnetic radiation.

The infrared radiation emitted by animals that is detectable with an infrared camera, and the cosmic microwave background radiation, are all examples of thermal radiation.

If a radiation-emitting object meets the physical characteristics of a black body in thermodynamic equilibrium, the radiation is called blackbody radiation. Planck's law describes the spectrum of blackbody radiation, which depends solely on the object's temperature. Wien's displacement law determines the most likely frequency of the emitted radiation, and the Stefan–Boltzmann law gives the radiant intensity.

Thermal radiation is also one of the fundamental mechanisms of heat transfer.

Thermal radiation, also known as heat, is the emission of electromagnetic waves from all matter that has a temperature greater than absolute zero. It represents the conversion of thermal energy into electromagnetic energy. Thermal energy consists of the kinetic energy of random movements of atoms and molecules in matter. All matter with a temperature by definition is composed of particles which have kinetic energy, and which interact with each other. These atoms and molecules are composed of charged particles, i.e., protons and electrons, and kinetic interactions among matter particles result in charge-acceleration and dipole-oscillation. This results in the electrodynamic generation of coupled electric and magnetic fields, resulting in the emission of photons, radiating energy away from the body through its surface boundary. Electromagnetic radiation, including light, does not require the presence of matter to propagate and travels in the vacuum of space infinitely far if unobstructed.

The characteristics of thermal radiation depend on various properties of the surface it is emanating from, including its temperature, its spectral absorptivity and spectral emissive power, as expressed by Kirchhoff's law. The radiation is not monochromatic, i.e., it does not consist of just a single frequency, but comprises a continuous dispersion of photon energies, its characteristic spectrum. If the radiating body and its surface are in thermodynamic equilibrium and the surface has perfect absorptivity at all wavelengths, it is characterized as a black body. A black body is also a perfect emitter. The radiation of such perfect emitters is called black-body radiation. The ratio of any body's emission relative to that of a black body is the body's emissivity, so that a black body has an emissivity of unity.

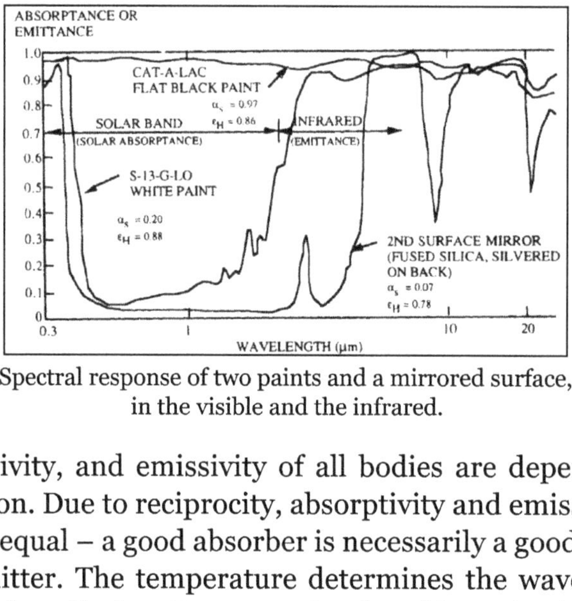

Spectral response of two paints and a mirrored surface, in the visible and the infrared.

Absorptivity, reflectivity, and emissivity of all bodies are dependent on the wavelength of the radiation. Due to reciprocity, absorptivity and emissivity for any particular wavelength are equal – a good absorber is necessarily a good emitter, and a poor absorber a poor emitter. The temperature determines the wavelength distribution of the electromagnetic radiation. For example, the white paint in the diagram to the right is highly reflective to visible light (reflectivity about 0.80), and so appears white to the human eye due to reflecting sunlight, which has a peak wavelength of about 0.5 micrometers. However, its emissivity at a temperature of about −5 °C (23 °F), peak wavelength of about 12 micrometers, is 0.95. Thus, to thermal radiation it appears black.

The distribution of power that a black body emits with varying frequency is described by Planck's law. At any given temperature, there is a frequency f_{max} at which the power emitted is a maximum. Wien's displacement law, and the fact that the frequency is inversely proportional to the wavelength, indicates that the peak frequency f_{max} is proportional to the absolute temperature T of the black body. The photosphere of the sun, at a temperature of approximately 6000 K, emits radiation principally in the (humanly) visible portion of the electromagnetic spectrum. Earth's atmosphere is partly transparent to visible light, and the light reaching the surface is absorbed or reflected. Earth's surface emits the absorbed radiation, approximating the behavior of a black body at 300 K with spectral peak at f_{max}. At these lower frequencies, the atmosphere is largely opaque and radiation from Earth's surface is absorbed or scattered by the atmosphere. Though about 10% of this radiation escapes into space, most is absorbed and then re-emitted by atmospheric gases. It is this spectral selectivity of the atmosphere that is responsible for the planetary greenhouse effect, contributing to global warming and climate change in general (but also critically contributing to climate stability when the composition and properties of the atmosphere are not changing).

The incandescent light bulb has a spectrum overlapping the black body spectra of the sun and the earth. Some of the photons emitted by a tungsten light bulb filament at 3000 K are in the visible spectrum. Most of the energy is associated with photons of longer wavelengths; these do not help a person see, but still transfer heat to the environment, as can be deduced empirically by observing an incandescent light bulb. Whenever EM radiation is emitted and then absorbed, heat is transferred. This principle is used in microwave ovens, laser cutting, and RF hair removal.

Unlike conductive and convective forms of heat transfer, thermal radiation can be concentrated in a tiny spot by using reflecting mirrors. Concentrating solar power takes advantage of this fact. In many such systems, mirrors are employed to concentrate sunlight into a smaller area. Instead of mirrors, Fresnel lenses can also be used to concentrate heat flux. (In principle, any kind of lens can be used, but only the Fresnel lens design is practical for very large lenses.) Either method can be used to quickly vaporize water into steam using sunlight. For example, the sunlight reflected from mirrors heats the PS10 Solar Power Plant, and during the day it can heat water to 285 °C (558.15 K) or 545 °F.

Surface Effects

Lighter colors and also whites and metallic substances absorb less illuminating light, and thus heat up less; but otherwise color makes small difference as regards heat transfer between an object at everyday temperatures and its surroundings, since the dominant emitted wavelengths are nowhere near the visible spectrum, but rather in the far infrared. Emissivities at those wavelengths have little to do with visual emissivities (visible colors); in the far infra-red, most objects have high emissivities. Thus, except in sunlight, the color of clothing makes little difference as regards warmth; likewise, paint color of houses makes little difference to warmth except when the painted part is sunlit.

The main exception to this is shiny metal surfaces, which have low emissivities both in the visible wavelengths and in the far infrared. Such surfaces can be used to reduce heat transfer in both directions; an example of this is the multi-layer insulation used to insulate spacecraft.

Low-emissivity windows in houses are a more complicated technology, since they must have low emissivity at thermal wavelengths while remaining transparent to visible light.

Nanostructures with spectrally selective thermal emittance properties offer numerous technological applications for energy generation and efficiency, e.g., for cooling photovoltaic cells and buildings. These applications require high emittance in the frequency range corresponding to the atmospheric transparency window in 8 to 13 micron wavelength range. A selective emitter radiating strongly in this range is thus exposed to the clear sky, enabling the use of the outer space as a very low temperature heat sink.

Personalized cooling technology is another example of an application where optical spectral selectivity can be beneficial. Conventional personal cooling is typically achieved through heat conduction and convection. However, the human body is a very efficient emitter of infrared radiation, which provides an additional cooling mechanism. Most conventional fabrics are opaque to infrared radiation and block thermal emission from the body to the environment. Fabrics for personalized cooling applications have been proposed that enable infrared transmission to directly pass through clothing, while being opaque at visible wavelengths. Fabrics that are transparent in the infrared can radiate body heat at rates that will significantly reduce the burden on power-hungry air-conditioning systems.

Properties

There are four main properties that characterize thermal radiation (in the limit of the far field):

- Thermal radiation emitted by a body at any temperature consists of a wide range of frequencies. The frequency distribution is given by Planck's law of black-body radiation for an idealized emitter as shown in the diagram at top.
- The dominant frequency (or color) range of the emitted radiation shifts to higher frequencies as the temperature of the emitter increases. For example, a *red hot* object radiates mainly in the long wavelengths (red and orange) of the visible band. If it is heated further, it also begins to emit discernible amounts of green and blue light, and the spread of frequencies in the entire visible range cause it to appear white to the human eye; it is *white hot*. Even at a white-hot temperature of 2000 K, 99% of the energy of the radiation is still in the infrared. This is determined by Wien's displacement law. In the diagram the peak value for each curve moves to the left as the temperature increases.
- The total amount of radiation of all frequencies increases steeply as the temperature rises; it grows as T^4, where T is the absolute temperature of the body. An object at the temperature of a kitchen oven, about twice the room temperature on the absolute temperature scale (600 K vs. 300 K) radiates 16 times as

much power per unit area. An object at the temperature of the filament in an incandescent light bulb—roughly 3000 K, or 10 times room temperature—radiates 10,000 times as much energy per unit area. The total radiative intensity of a black body rises as the fourth power of the absolute temperature, as expressed by the Stefan–Boltzmann law. In the plot, the area under each curve grows rapidly as the temperature increases.

- The rate of electromagnetic radiation emitted at a given frequency is proportional to the amount of absorption that it would experience by the source, a property known as reciprocity. Thus, a surface that absorbs more red light thermally radiates more red light. This principle applies to all properties of the wave, including wavelength (color), direction, polarization, and even coherence, so that it is quite possible to have thermal radiation which is polarized, coherent, and directional, though polarized and coherent forms are fairly rare in nature far from sources (in terms of wavelength).

Near-field and Far-field

The general properties of thermal radiation as described by the Planck's law apply if the linear dimension of all parts considered, as well as radii of curvature of all surfaces are large compared with the wavelength of the ray considered' (typically from 8-25 micrometres for the emitter at 300 K). Indeed, thermal radiation as discussed above takes only radiating waves (far-field, or electromagnetic radiation) into account. A more sophisticated framework involving electromagnetic theory must be used for smaller distances from the thermal source or surface (near-field thermal radiation). For example, although far-field thermal radiation at distances from surfaces of more than one wavelength is generally not coherent to any extent, near-field thermal radiation (i.e., radiation at distances of a fraction of various radiation wavelengths) may exhibit a degree of both temporal and spatial coherence.

Planck's law of thermal radiation has been challenged in recent decades by predictions and successful demonstrations of the radiative heat transfer between objects separated by nanoscale gaps that deviate significantly from the law predictions. This deviation is especially strong (up to several orders in magnitude) when the emitter and absorber support surface polariton modes that can couple through the gap separating cold and hot objects. However, to take advantage of the surface-polariton-mediated near-field radiative heat transfer, the two objects need to be separated by ultra-narrow gaps on the order of microns or even nanometers. This limitation significantly complicates practical device designs.

Another way to modify the object thermal emission spectrum is by reducing the dimensionality of the emitter itself. This approach builds upon the concept of confining electrons in quantum wells, wires and dots, and tailors thermal emission by engineering confined photon states in two- and three-dimensional potential traps, including wells, wires, and dots. Such spatial confinement concentrates photon states and enhances thermal emission at select frequencies. To achieve the required level of photon confinement, the dimensions of the radiating objects should be on the order of or below the thermal wavelength predicted by Planck's law. Most importantly, the emission spectrum of thermal wells, wires and dots deviates from Planck's law predictions not only in the near field, but also in the far field, which significantly expands the range of their applications.

Table: Subjective color to the eye of a black body thermal radiator.

°C (°F)	Subjective color
480 °C (896 °F)	faint red glow
580 °C (1,076 °F)	dark red
730 °C (1,350 °F)	bright red, slightly orange
930 °C (1,710 °F)	bright orange
1,100 °C (2,010 °F)	pale yellowish orange
1,300 °C (2,370 °F)	yellowish white
> 1,400 °C (2,550 °F)	white (yellowish if seen from a distance through atmosphere)

Selected Radiant Heat Fluxes

The time to a damage from exposure to radiative heat is a function of the rate of delivery of the heat. Radiative heat flux and effects: (1 W/cm^2 = 10 kW/m^2).

kW/m^2	Effect
170	Maximum flux measured in a post-flashover compartment
80	Thermal Protective Performance test for personal protective equipment
52	Fiberboard ignites at 5 seconds
29	Wood ignites, given time
20	Typical beginning of flashover at floor level of a residential room
16	Human skin: sudden pain and second-degree burn blisters after 5 seconds
12.5	Wood produces ignitable volatiles by pyrolysis
10.4	Human skin: Pain after 3 seconds, second-degree burn blisters after 9 seconds
6.4	Human skin: second-degree burn blisters after 18 seconds
4.5	Human skin: second-degree burn blisters after 30 seconds
2.5	Human skin: burns after prolonged exposure, radiant flux exposure typically encountered during firefighting
1.4	Sunlight, sunburns potentially within 30 minutes. Note sunburn is NOT a thermal burn, it is caused by DNA damage due to ultraviolet radiation.

Interchange of Energy

Radiant heat panel for testing precisely quantified energy exposures at National Research Council, near Ottawa, Ontario, Canada.

Thermal radiation is one of the three principal mechanisms of heat transfer. It entails the emission of a spectrum of electromagnetic radiation due to an object's temperature. Other mechanisms are convection and conduction. The interplay of energy exchange by thermal radiation is characterized by the following equation:

$$\alpha + \rho + \tau = 1.$$

Here, α represents the spectral absorption component, ρ spectral reflection component and τ the spectral transmission component. These elements are a function of the wavelength (λ) of the electromagnetic radiation. The spectral absorption is equal to the emissivity ϵ ; this relation is known as Kirchhoff's law of thermal radiation. An object is called a black body if, for all frequencies, the following formula applies:

$$\alpha = \epsilon = 1.$$

In a practical situation and room-temperature setting, humans lose considerable energy due to thermal radiation in infra-red in addition to that lost by conduction to air (aided by concurrent convection, or other air movement like drafts). The heat energy lost is partially regained by absorbing heat radiation from walls or other surroundings. (Heat gained by conduction would occur for air temperature higher than body temperature.) Otherwise, body temperature is maintained from generated heat through internal metabolism. Human skin has an emissivity of very close to 1.0. Using the formulas below shows a human, having roughly 2 square meter in surface area, and a temperature of about 307 K, continuously radiates approximately 1000 watts. If people are indoors, surrounded by surfaces at 296 K, they receive back about 900 watts from the wall, ceiling, and other surroundings, so the net loss is only about 100 watts. These heat transfer estimates are highly dependent on extrinsic variables, such as wearing clothes, i.e. decreasing total thermal circuit conductivity, therefore reducing total output heat flux. Only truly *gray* systems (relative equivalent emissivity/absorptivity and no directional transmissivity dependence in *all* control volume bodies considered) can achieve reasonable steady-state heat flux estimates through the Stefan-Boltzmann law. Encountering this "ideally calculable" situation is almost impossible (although common engineering procedures surrender the dependency of these unknown variables and "assume" this to be the case). Optimistically, these "gray" approximations will get close to real solutions, as most divergence from Stefan-Boltzmann solutions is very small (especially in most STP lab controlled environments).

If objects appear white (reflective in the visual spectrum), they are not necessarily equally reflective (and thus non-emissive) in the thermal infrared – see the diagram at the left. Most household radiators are painted white, which is sensible given that they are not hot enough to radiate any significant amount of heat, and are not designed as thermal radiators at all – instead, they are actually convectors, and painting them matt black would make little difference to their efficacy. Acrylic and urethane based white paints have 93% blackbody radiation efficiency at room temperature (meaning the term "black body" does not always correspond to the visually perceived color of an object). These materials that do not follow the "black color = high emissivity/absorptivity" caveat will most likely have functional spectral emissivity/absorptivity dependence.

Calculation of radiative heat transfer between groups of object, including a 'cavity' or 'surroundings' requires solution of a set of simultaneous equations using the radiosity method. In these calculations, the geometrical configuration of the problem is distilled to a set of numbers called view factors, which give the proportion of radiation leaving any given surface that hits another specific surface. These calculations are important in the fields of solar thermal energy, boiler and furnace design and raytraced computer graphics.

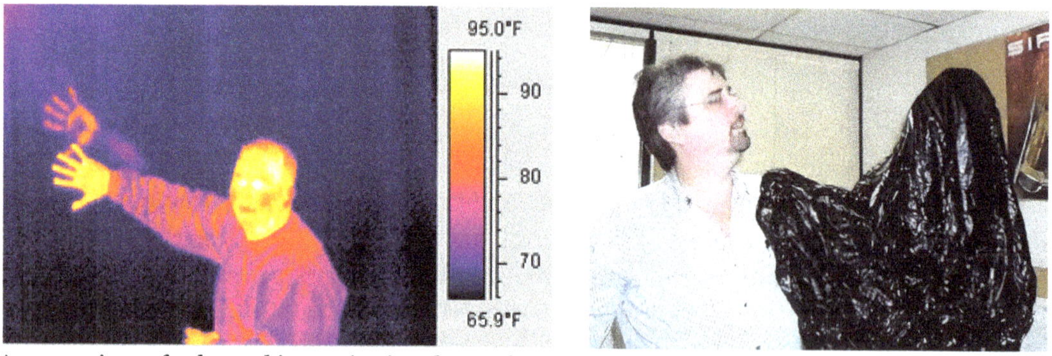

A comparison of a thermal image (top) and an ordinary photograph (bottom) shows that a trash bag is transparent, but glass (the man's spectacles) is opaque in long-wavelength infrared.

A selective surface can be used when energy is being extracted from the sun. For instance, when a green house is made, most of the roof and walls are made out of glass. Glass is transparent in the visible (approximately 0.4 μm < λ < 0.8 μm) and near-infrared wavelengths, but opaque to mid- to far-wavelength infrared (approximately λ > 3 μm). Therefore, glass lets in radiation in the visible range, allowing us to be able to see through it, but does not let out radiation that is emitted from objects at or close to room temperature. This traps what we feel as heat. This is known as the greenhouse effect and can be observed by getting into a car that has been sitting in the sun.Selective surfaces can also be used on solar collectors. We can find out how much help a selective surface coating is by looking at the equilibrium temperature of a plate that is being heated through solar radiation. If the plate is receiving a solar irradiation of 1350 W/m² (minimum is 1325 W/m² on July 4 and maximum is 1418 W/m² on January 3) from the sun the temperature of the plate where the radiation leaving is equal to the radiation being received by the plate is 393 K (248 °F). If the plate has a selective surface with an emissivity of 0.9 and a cut off wavelength of 2.0 μm, the equilibrium temperature is approximately 1250 K (1790 °F). The calculations were made neglecting convective heat transfer and neglecting the solar irradiation absorbed in the clouds/ atmosphere for simplicity, the theory is still the same for an actual problem.

To reduce the heat transfer from a surface, such as a glass window, a clear reflective film with a low emissivity coating can be placed on the interior of the surface. "Low-emittance (low-E) coatings are microscopically thin, virtually invisible, metal or metallic oxide layers deposited on a window or skylight glazing surface primarily to reduce the U-factor by suppressing radiative heat flow". By adding this coating we are limiting the amount of radiation that leaves the window thus increasing the amount of heat that is retained inside the window.

Since any electromagnetic radiation, including thermal radiation, conveys momentum as well as energy, thermal radiation also induces very small forces on the radiating or absorbing objects. Normally these forces are negligible, but they must be taken into account when considering spacecraft navigation. The Pioneer anomaly, where the motion of the craft slightly deviated from that expected from gravity alone, was eventually tracked down to asymmetric thermal radiation from the spacecraft. Similarly, the orbits of asteroids are perturbed since the asteroid absorbs solar radiation on the side facing the sun, but then re-emits the energy at a different angle as the rotation of the asteroid carries the warm surface out of the sun's view (the YORP effect).

Radiative Power

Thermal radiation power of a black body per unit area of radiating surface per unit of solid angle and per unit frequency υ is given by Planck's law as:

$$u(v,T) = \frac{2hv^3}{c^2} \cdot \frac{1}{e^{hv/k_B T} - 1}$$

or instead of per unit frequency, per unit wavelength as,

$$u(\lambda,T)=\frac{2hc^2}{\lambda^5}\cdot\frac{1}{e^{hc/k_BT\lambda}-1}$$

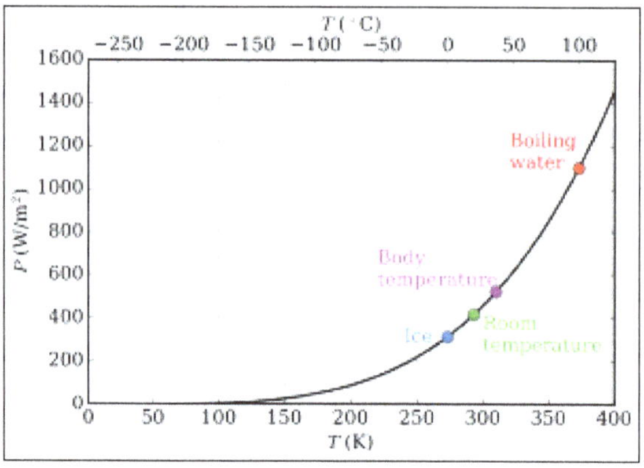

Power emitted by a black body plotted against the temperature according to the Stefan–Boltzmann law.

This formula mathematically follows from calculation of spectral distribution of energy in quantized electromagnetic field which is in complete thermal equilibrium with the radiating object. Plancks law shows that radiative energy increases with temperature, and explains why the peak of an emission spectrum shifts to shorter wavelengths at higher temperatures. It can also be found that energy emitted at shorter wavelengths increases more rapidly with temperature relative to longer wavelengths. The equation is derived as an infinite sum over all possible frequencies in a semi-sphere region. The energy, $E=hv$,, of each photon is multiplied by the number of states available at that frequency, and the probability that each of those states will be occupied.

Integrating the above equation over v the power output given by the Stefan–Boltzmann law is obtained, as:

$$P=\sigma\cdot A\cdot T^4$$

where the constant of proportionality σ is the Stefan–Boltzmann constant and A is the radiating surface area.

The wavelength λ, for which the emission intensity is highest, is given by Wien's displacement law as:

$$\lambda_{\max}=\frac{b}{T}$$

For surfaces which are not black bodies, one has to consider the (generally frequency dependent) emissivity factor $\epsilon(v)$. This factor has to be multiplied with the radiation spectrum formula before integration. If it is taken as a constant, the resulting formula for the power output can be written in a way that contains ò as a factor:

$$P = \epsilon \cdot \sigma \cdot A \cdot T^4$$

This type of theoretical model, with frequency-independent emissivity lower than that of a perfect black body, is often known as a *grey body*. For frequency-dependent emissivity, the solution for the integrated power depends on the functional form of the dependence, though in general there is no simple expression for it. Practically speaking, if the emissivity of the body is roughly constant around the peak emission wavelength, the gray body model tends to work fairly well since the weight of the curve around the peak emission tends to dominate the integral.

Constants

Definitions of constants used in the above equations:

h	Planck's constant	$6.626\ 069\ 3(11)\times10^{-34}$ J·s = $4.135\ 667\ 43(35)\times10^{-15}$ eV·s
b	Wien's displacement constant	$2.897\ 768\ 5(51)\times10^{-3}$ m·K
k_B	Boltzmann constant	$1.380\ 650\ 5(24)\times10^{-23}$ J·K^{-1} = $8.617\ 343\ (15)\times10^{-5}$ eV·K^{-1}
σ	Stefan–Boltzmann constant	$5.670\ 373\ (21)\times10^{-8}$ W·m^{-2}·K^{-4}
c	Speed of light	299 792 458 m·s^{-1}

Variables

Definitions of variables, with example values:

T	Absolute temperature	For units used above, must be in kelvins (e.g. average surface temperature on Earth = 288 K)
A	Surface area	$A_{cuboid} = 2ab + 2bc + 2ac$; $A_{cylinder} = 2\pi \cdot r(h + r)$; $A_{sphere} = 4\pi \cdot r^2$

Radiative Heat Transfer

The *net* radiative heat transfer from one surface to another is the radiation leaving the first surface for the other minus that arriving from the second surface.

- For black bodies, the rate of energy transfer from surface 1 to surface 2 is:

$$\dot{Q}_{1\to2} = A_1 E_{b1} F_{1\to2} - A_2 E_{b2} F_{2\to1}$$

where σ is surface area, E_b is energy flux (the rate of emission per unit surface area) and $F_{1\to2}$ is the view factor from surface 1 to surface 2. Applying both the reciprocity rule for view factors, $A_1F_{1\to2} = A_2F_{2\to1}$, and the Stefan–Boltzmann law, $E_b = \sigma T^4$, yields:

$$\dot{Q}_{1\to2} = \sigma A_1 F_{1\to2}(T_1^4 - T_2^4)$$

where is the Stefan–Boltzmann constant and T is temperature. A negative value for $\dot{Q}$ indicates that net radiation heat transfer is from surface 2 to surface 1.

- For two grey-body surfaces forming an enclosure, the heat transfer rate is:

$$\dot{Q} = \frac{\sigma(T_1^4 - T_2^4)}{\frac{1-\epsilon_1}{A_1\epsilon_1} + \frac{1}{A_1F_{1\to2}} + \frac{1-\epsilon_2}{A_2\epsilon_2}}$$

where ϵ_1 and ϵ_2 are the emissivities of the surfaces

Formulas for radiative heat transfer can be derived for more particular or more elaborate physical arrangements, such as between parallel plates, concentric spheres and the internal surfaces of a cylinder.

RADIATION PROPERTIES

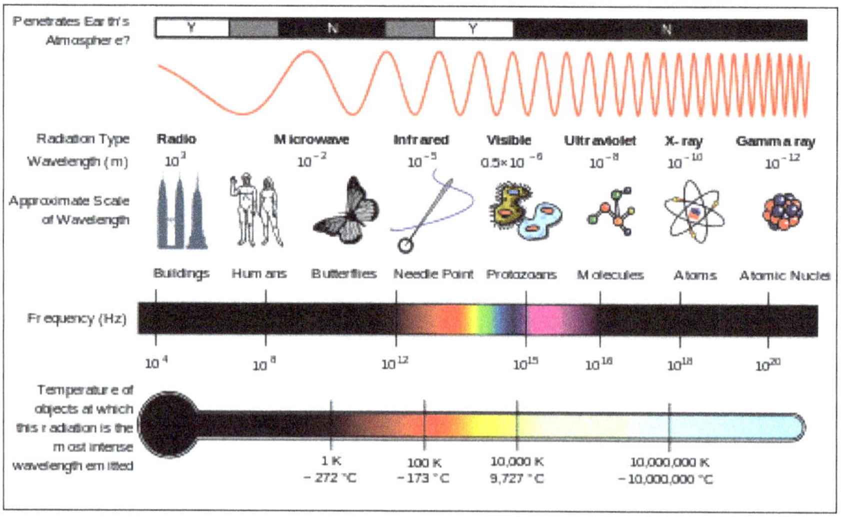

Schematic of the electromagnetic spectrum.

Conduction and convection are heat transfer processes that require the presence of a medium. Radiation heat transfer is characteristically different from the other two in that it does not require a medium and, in fact it reaches maximum efficiency in a vacuum. Electromagnetic radiation has some proper characteristics depending on the frequency and wavelengths of the radiation. The phenomenon of radiation is not yet fully understood. Two theories have been used to explain radiation; however neither of them is perfectly satisfactory.

First, the earlier theory which originated from the concept of a hypothetical medium referred as ether. Ether supposedly fills all evacuated or non evacuated spaces. The transmission of light or of radiant heat are allowed by the propagation of electromagnetic waves in the ether. Electromagnetic waves have similar characteristics to television and radio broadcasting waves they only differ in wavelength. All electromagnetic waves travel at the same speed; therefore, shorter wavelengths are associated with high frequencies. Since every body or fluid is submerged in the ether, due to the vibration of the molecules, any body or fluid can potentially initiate an electromagnetic wave. All bodies generate and receive electromagnetic waves at the expense of its stored energy The second theory of radiation is best known as the quantum theory and was first offered by Max Planck in 1900. According to this theory, energy emitted by a radiator is not continuous but is in the form of quanta. Planck claimed that quantities had different sizes and frequencies of vibration similarly to the wave theory. The energy E is found by the expression $E = hv$, where h is the Planck's constant and v is the frequency. Higher frequencies are originated by high temperatures and create an increase of energy in the quantum. While the propagation of electromagnetic waves of all wavelengths is often referred as "radiation," thermal radiation is often constrained to the visible and infrared regions. For engineering purposes, it may be stated that thermal radiation is a form of electromagnetic radiation which varies on the nature of a surface and its temperature. Radiation waves may travel in unusual patterns compared to conduction heat flow. Radiation allows waves to travel from a heated body through a cold nonabsorbing or partially absorbing medium and reach a warmer body again. This is the case of the radiation waves that travel from the sun to the earth.

Properties

Emissivity (ε)

The emissivity of a given surface is the measure of its ability to emit radiation energy in comparison to a blackbody at the same temperature. The emissivity of a surface varies between zero and one. The emissivity of a real surface varies as a function of the surface temperature, the wavelength, and the direction of the emitted radiation. The fundamental emissivity of a surface at a given temperature is the spectral directional emissivity, which is defined as the ratio of the intensity of radiation emitted by the surface at a specified wavelength and direction to that emitted by a blackbody under the same conditions. The total directional emissivity is defined in the same fashion by using the total intensities integrated over all wavelengths. In practice, a more convenient method is used: hemispherical properties. These properties are spectrally and directionally averaged. The emissivity of a surface at a specified wavelength may vary as temperature changes since the spectral distribution of emitted radiation changes with temperature. Finally the total hemispherical emissivity is defined in terms of the radiation energy emitted over all wavelengths in all directions. Radiation is a complex phenomenon. The dependability of its properties in wavelength and direction makes it even more complicated. Therefore, the gray and diffuse approximation methods are commonly used to perform radiation calculations. A gray surface is characterized by having properties independent of wavelength, and a diffuse surface has properties independent of direction.

Absorptivity (α), Reflectivity (ρ) and Transmissivity (t)

If the amounts of radiation energy absorbed, reflected, and transmitted when radiation strikes a surface are measured in percentage of the total energy in the incident electromagnetic waves, the total energy will be divided into three groups. They are called absorptivity (α), reflectivity (ρ) and transmissivity (t):

$$\alpha + \rho + t = 1 \quad (1)$$

- Absorption is the fraction of radiation absorbed by a surface.
- Reflectivity is the fraction reflected by the surface.
- Transmissivity is the fraction transmitted by the surface.

A body is considered transparent if it can transmit some of the radiation waves falling on its surface. If electromagnetic waves are not transmitted through the substance it is therefore called opaque. When radiation waves hit the surface of an opaque body, some of the waves are reflected back while the other waves are absorbed by a thin layer of the material close to the surface. For opaque bodies, Equation 1 reduces to:

$$\alpha + \rho = 1 \quad (2)$$

Reflectivity deviates from the other properties in that it is bidirectional in nature. In other words, this property depends on the direction of the incident of radiation as well as the direction of the reflection. Therefore, the reflected rays of a radiation spectrum incident on a real surface in a specified direction forms an irregular shape that is not easily predictable. In practice, surfaces are assumed to reflect in a perfectly specular or diffuse manner. In a specular reflection, the angles of reflection and incidence are equal. In diffuse reflection, radiation is reflected equally in all directions. Reflection from smooth and polished surfaces can be assumed to be specular reflection, whereas reflection from rough surfaces approximates diffuse reflection. In radiation analysis a surface is defined as smooth if the height of the surface roughness is much smaller relative to the wavelength of the incident radiation.

RADIATION INTENSITY

In radiometry, radiant intensity is the radiant flux emitted, reflected, transmitted or received, per unit solid angle, and spectral intensity is the radiant intensity per unit frequency or wavelength, depending on whether the spectrum is taken as a function of frequency or of wavelength. These are *directional* quantities. The SI unit of radiant intensity is the watt per steradian (W/sr), while that of spectral intensity in frequency is the watt per steradian per hertz ($W \cdot sr^{-1} \cdot Hz^{-1}$) and that of spectral intensity in wavelength is the watt per steradian per metre ($W \cdot sr^{-1} \cdot m^{-1}$)—commonly the watt per steradian per nanometre ($W \cdot sr^{-1} \cdot nm^{-1}$). Radiant intensity is distinct from irradiance and radiant exitance, which are often called *intensity* in branches of physics other than radiometry. In radio-frequency engineering, radiant intensity is sometimes called radiation intensity.

Mathematical Definitions

Radiant Intensity

Radiant intensity, denoted $I_{e,\Omega}$ ("e" for "energetic", to avoid confusion with photometric quantities, and "Ω" to indicate this is a *directional* quantity), is defined as,

$$I_{e,\Omega} = \frac{\partial \Phi_e}{\partial \Omega},$$

where,

- ∂ is the partial derivative symbol.
- Φ_e is the radiant flux emitted, reflected, transmitted or received.
- Ω is the solid angle.

In general, $I_{e,\Omega}$ is a function of viewing angle θ and potentially azimuth angle. For the special case of a Lambertian surface, $I_{e,\Omega}$ follows the Lambert's cosine law $I_{e,\Omega} = I_o \cos \theta$.

When calculating the radiant intensity emitted by a source, Ω refers to the solid angle into which the light is emitted. When calculating radiance received by a detector, Ω refers to the solid angle subtended by the source as viewed from that detector.

Spectral Intensity

Spectral intensity in frequency, denoted $I_{e,\Omega,\nu}$, is defined as,

$$I_{e,\Omega,\nu} = \frac{\partial I_{e,\Omega}}{\partial \nu},$$

where, v is the frequency.

Spectral intensity in wavelength, denoted $I_{e,\Omega,\lambda}$, is defined as,

$$I_{e,\Omega,\lambda} = \frac{\partial I_{e,\Omega}}{\partial \lambda},$$

where λ is the wavelength.

Radio-frequency Engineering

Radiant intensity is used to characterize the emission of radiation by an antenna:

$$I_{e,\Omega} = E_e(r)r^2,$$

where,

- E_e is the irradiance of the antenna.
- r is the distance from the antenna.

Unlike power density, radiant intensity does not depend on distance: because radiant intensity is defined as the power through a solid angle, the decreasing power density over distance due to the inverse-square law is offset by the increase in area with distance.

SHAPE FACTOR ALGEBRA FOR RADIATION EXCHANGE

There is a factor called the SHAPE FACTOR ($F_{ij)}$ OR view factor or configuration factor or form factor or geometric factor which accounts for the orientation and geometry of the surfaces exchanging radiations. Shape factor F12 = ratio of radiations intercepted by surface 2 / radiations leaving surface 1. Its symbol is F_{ij}. Say for example 140 units of energy emitted by body A. 35 units of energy are received by a body B. Then F_{AB}= 35/140=0.25. It depends how the surfaces view each other and also how the surfaces radiate energy with respect to each other.

Laws for Shape Factors

- Reciprocity law $A_iF_{ij} = A_jF_{ji}$ OR $A_2\ F_{23} = A_3\ F_{32}$

  Say surface 1 is convex and is enclosed in another surface 2 (a small ball in a big ball), then F12=1 and from the law of reciprocity F21=A_1/A_2

- Summation law – Say for four bodies

a. $F_{11}+F_{12}+F_{13}+F_{14}=1$

b. $F_{21}+F_{22}+F_{23}+F_{24}=1$

c. $F_{31}+F_{32}+F_{33}+F_{34}=1$

d. $F_{41}+F_{42}+F_{43}+F_{44}=1$

- Shape factor for a self body:

  F_{ii} or F_{11} energy received by body 1 out of energy emitted by body 1

  Shape factor for a flat surface $F_{ii}=0$

  Further, for a convex surface $F_{ii}=0$ and hence $F_{11}=F_{22}=F_{33}=0$

  For a concave surface $F_{ii}\neq 0$ and hence $F_{11}\neq 0$, $F_{22}\neq 0$, $F_{33}\neq 0$

- If there are 'n' surfaces, there are n^2 shape factors:

  For example for three bodies, 9 shape factors are required.

How to Find these n^2 Shape Factors.

There are three steps:

- Apply summation rule to get n equations to give n shape factors i.e. 3
- Use Reciprocity relation $n(n-1)/2$ times to get $n(n-1)/2$ shape factors i.e. 3
- Now it is required to find the remaining shape factors i.e. Remaining= $n^2-n-n(n-1)/2$ shape factors i.e. 3 in this case.

For three surfaces,

Remaining shape factors will be = $3^2-3-3(3-1)/2=9-3-3=03$.

These remaining shape factors are found from general observations like $F_{11}=0$, $F_{12}=1$, $F_{21}=A_1/A_2$ etc.

Shape Factor Algebra

Equations for shape factors are not available for complex shapes. Complex shapes are converted into simpler shapes. Shape factors then for simpler shapes can be found from the shape factor algebra (Laws of shape factors) which is based on:

- The definition of shape factor.
- Law of reciprocity.
- Summation law

Say there are two surfaces A_1 and A_2. Say receiving surface A_2 is a complex surface. It is divided further into two areas A_3 and A_4 in simple areas.

$$A_1F_{12}=A_1F_{13}+A_1F_{14}$$

Shape factors can be found from standard graphs for the following cases:

- Parallel surfaces (rectangular, equal circular, non-equal circular, etc).
- For surfaces at right angles.

GREY BODIES HEAT EXCHANGE

The easiest method to calculate radiative heat transfer between two bodies is when they are assumed to be black bodies. However, in reality most surface are grey bodies. Recall that grey bodies absorb a certain amount of radiation while reflecting a portion of the radiation off of the surface back into space.

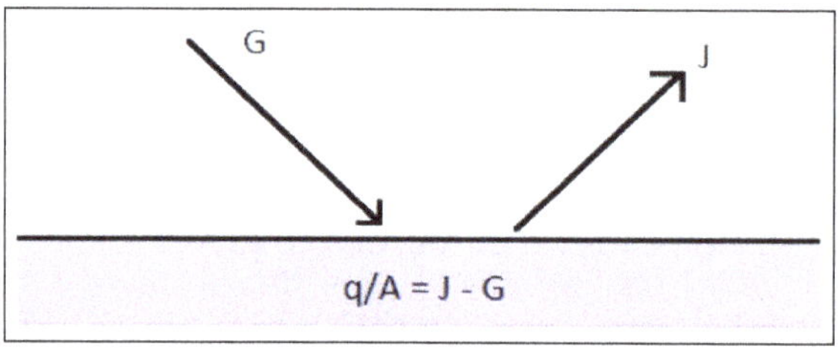

In the above image G represent irradiation which is the total radiation that come in contact with a surface per unit time and unit area. While J represents the radiosity which is the total amount of radiation that is reflected off a surface per unit time and unit area. The equation below can be used to determine the value for J.

$$J=\varepsilon E_b+(1-\varepsilon)G$$

ε = The Emissivity of the Object.

E_b The Energy Emitted from a Black Body.

To determine the net energy leaving the surface you would need to find the difference between the radiosity and irradiation as seen in the equation below:

$$\frac{q}{A}=\varepsilon E_b+(1-\varepsilon)G-G$$

A = Surface Area.

Solving equation 1 for G and implementing it into equation 2 will than yield the following equation.

$$q = \frac{E_b - J}{(1-\varepsilon)/\varepsilon A}$$

Finally, the radiative heat transfer for a grey body can be represented as a resistance as seen from the equation below.

$$R = \frac{(1-\varepsilon)}{\varepsilon A}$$

Equation shows you how to calculate the total radiative heat transfer when thermal radiation strikes a grey body. It however does not consider how different grey bodies will react with each other. When dealing with multiple bodies you will have to determine the radiation shape factor. Refer to the equation below to see the heat transfer between two bodies when the radiation shape factor is considered.

$$q_{1-2} = \frac{J_1 - J_2}{1/A_1F_{1-2}}$$

F_{1-2} = The Shape Factor between surface 1 and 2.

The equation to determine the resistance to radiative heat transfer for the radiation shape factor can be seen below.

$$R = \frac{1}{A_1F_{1-2}}$$

Now what if you have two surfaces that are grey bodies that see only each other and nothing else? The image below represents the thermal resistances within the system.

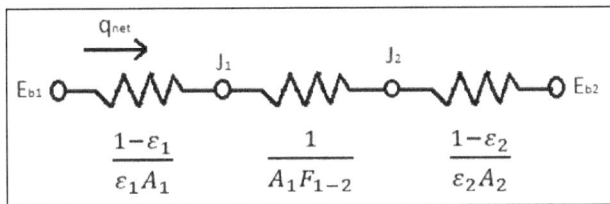

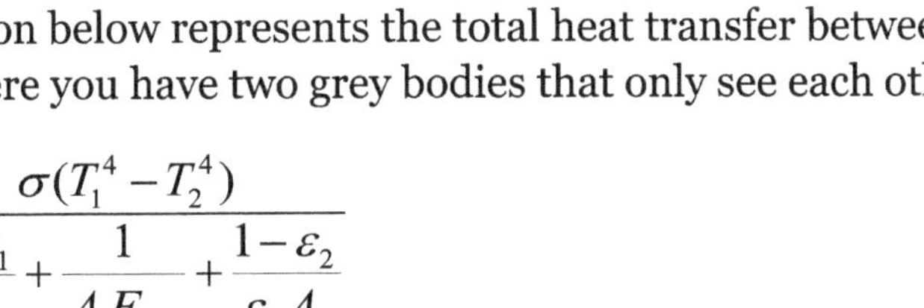

Finally, the equation below represents the total heat transfer between the two surfaces for the system where you have two grey bodies that only see each other.

$$q_{net} = \frac{\sigma(T_1^4 - T_2^4)}{\frac{1-\varepsilon_1}{\varepsilon_1 A_1} + \frac{1}{A_1F_{1-2}} + \frac{1-\varepsilon_2}{\varepsilon_2 A_2}}$$

$$\sigma = 5.669 \text{ X } 10^{-8} \text{ W/m}^2\text{-K}^4$$

Now what if you have two grey bodies radiating heat between each other and also radiating heat into a large room? This problem would be represented by the resistance diagram below.

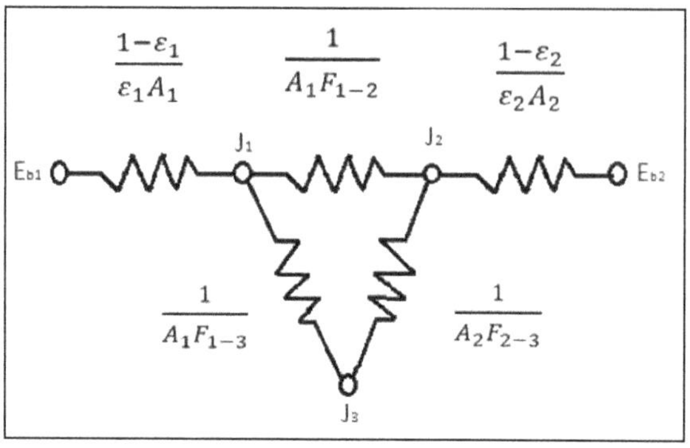

$$E_{b1} = \sigma T_1^4$$

$$E_{b2} = \sigma T_2^4$$

$$E_{b3} = \sigma T_3^4 = J3$$

To determine the radiositie at J_1 and J_2 the following equations would be used.

$$\frac{E_{b1} - J_1}{(1-\varepsilon_1)/\varepsilon_1 A_1} + \frac{J_2 - J_1}{1/A_1 F_{1-2}} + \frac{J_3 - J_1}{1/A_1 F_{1-3}} = 0$$

$$\frac{J_1 - J_2}{1/A_1 F_{1-2}} + \frac{J_3 - J_2}{1/A_2 F_{2-3}} + \frac{E_{b2} - J_2}{(1-\varepsilon_2)/\varepsilon_2 A_2} = 0$$

Next to find the total heat loss from object 1 the following equation would be used.

$$q_1 = \frac{E_{b1} - J_1}{(1-\varepsilon_1)/\varepsilon_1 A_1}$$

To find the total heat loss for object 2 the following equation would be used.

$$q_2 = \frac{E_{b2} - J_2}{(1-\varepsilon_2)/\varepsilon_2 A_2}$$

To find the total heat transfer received by the room the following equation would be used.

$$q_3 = \frac{J_1 - J_3}{1/A_1 F_{1-3}} + \frac{J_2 - J_3}{1/A_2 F_{23}}$$

Finally, the net energy lost by the plates and then absorbed by the room would be represented by the following equation.

$$q_3 = q_1 + q_2$$

As more and more surfaces are considered the more complex the problem will become.

RADIATION SHIELDS

A radiation shield is a barrier wall of low emissivity placed between two surfaces which reduce the radiation between the bodies. In fact, the radiation shield will put additional resistance to the radiative heat transfer between the surfaces as shown in figure.

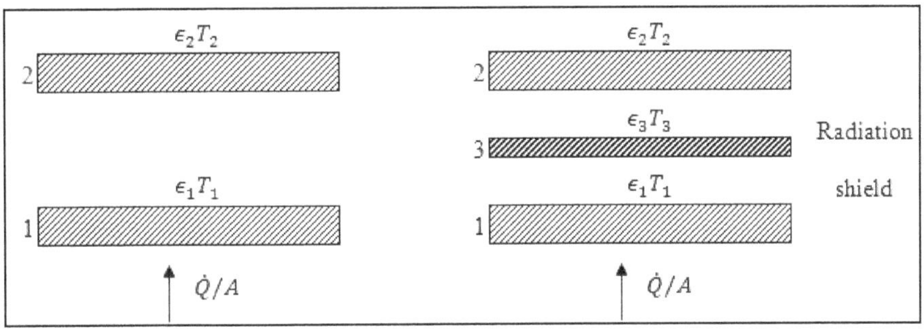

Radiation between two large infinite plates (a) without and (b) with radiation shield.

Considering figure and the system is at steady state, and the surfaces are flat (F_{ij} because each plate is in full view of the other). Moreover, the surface are large enough and $\frac{A_1}{A_2} \approx 1$ may be considered and the equivalent blackbody radiation energy may be written as $E_b = \sigma T^4$.

Thus,

$$\left.\frac{\dot{Q}}{A}\right|_{net} \left.\frac{\dot{Q}_{13}}{A_1}\right|_{net} \left.\frac{\dot{Q}_{32}}{A_3}\right|_{net} \frac{\sigma\left(T_1^4 - T_3^4\right)}{\frac{1}{\varepsilon_1} + \frac{1}{\varepsilon_3} - 1} = \frac{\sigma\left(T_3^4 - T_2^4\right)}{\frac{1}{\varepsilon_3} + \frac{1}{\varepsilon_2} - 1}$$

In order to have a feel of the role of the radiation shield, consider that the emissivities of all the three surfaces are equal.

$$\varepsilon_1 = \varepsilon_2 = \varepsilon_3 = \varepsilon$$

Then it can be seen that the heat flux is just one half of that which would be experienced if there were no shield present.

In similar line we can deduce that when n-shields are arranged between the two surfaces then,

$$\left(\frac{\dot{Q}}{A}\right)_{net\,with\,shield} = \frac{1}{n+1}\left(\frac{\dot{Q}}{A}\right)_{without\,shield}$$

Electrical network for radiation through absorbing and transmitting medium The previous discussions were based on the consideration that the heat transfer surfaces were separated by a completely transparent medium. However, in real situations the heat transfer medium absorbs as well as transmits. The examples of such medium are glass, plastic film, and various gases.

Consider two non-transmitting surfaces are separated by a transmitting and absorbing medium. The medium may be considered as a radiation shield which see themselves and others. If we distinguish the transparent medium by *m* and if the medium is non-reflective (say gas) then using Kirchhoff's law,

$$\alpha_m + \tau_m = 1 = \varepsilon_m + \tau_m$$

The energy leaving surface 1 which is transmitted through the medium and reaches the surface 2 is,

$$J_1\, A_1 F_{12}\, \tau_m$$

and that which leaves surface 2 and arrives at surface 1 is,

$$J_2\, A_2 F_{12}\, \tau_m$$

Therefore, the net exchange in the transmission process is therefore,

$$\dot{Q}_{12} = A_1 F_{12}\, \tau_m\, (J_1 - J_2)$$

Using $\alpha_m + \tau_m = 1 = \varepsilon_m + \tau_m$,

$$\dot{Q}_{12} = \frac{(J_1 - J_2)}{\left(\dfrac{1}{A_1 F_{12}^{(1-\epsilon m)}}\right)}$$

Thus the equivalent circuit diagram is shown in figure.

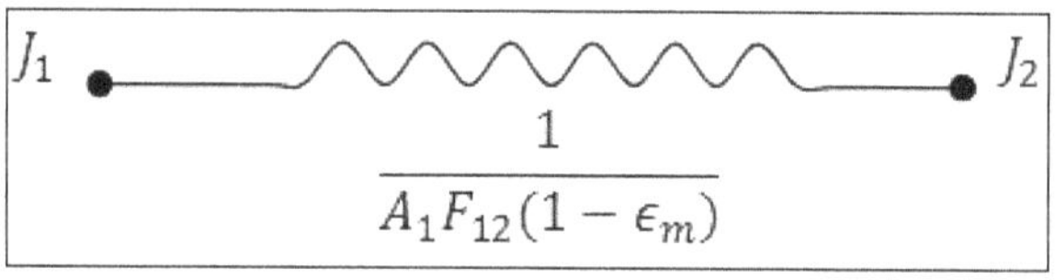

Equivalent electrical circuit for radiation through gas.

Radiation Combined with Conduction and Convection

In industrial processes, in general, the heat transfer at higher temperature has significant portion of radiation along with conduction and convection. For example, a heated surface is shown in the figure. with all the three mechanism of heat transfer.

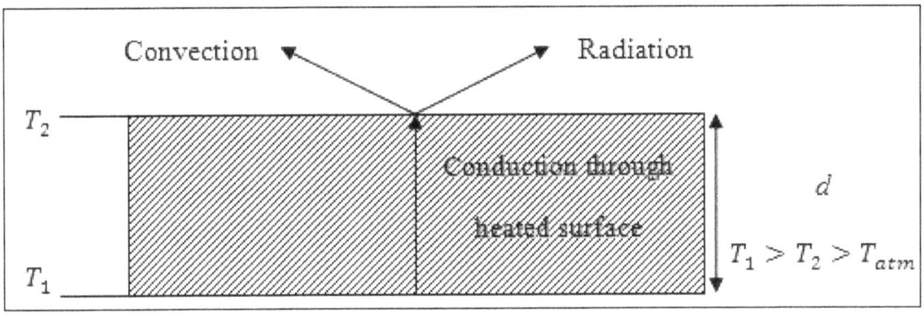

At steady state

Heat flux by conduction = heat flux by convention + heat flux by radiation,

$$\frac{k}{d}\left(T_1 - T_2\right) = h\left(T_2 - T_{atm}\right) + \in \sigma\left(T_2^4 - T_{atm}^4\right)$$

Where, h is the heat transfer coefficient at the surface in contact (outer surface) with atmosphere due to natural and forced convection combined together $\in$ is the emissivity of the outer surface, and T_{atm} is the atmospheric temperature.

References

- Black-body-radiation, 252, classes: virginia.edu, Retrieved 30 August, 2019
- “2018 codata value: wien wavelength displacement law constant”. The nist reference on constants, units, and uncertainty. Nist. 20 may 2019. Retrieved 2019-05-20
- Radiation-heat-transfer, kirchhoffs-law-of-thermal-radiation, heat-transfer, nuclear-engineering: nuclear-power.net, Retrieved 16 January, 2019
- Fan, shanhui; li, wei (11 june 2018). “nanophotonic control of thermal radiation for energy applications [invited]”. Optics express. 26(12): 15995–16021. Doi:10.1364/oe.26.015995. Issn 1094-4087
- Shape-factor-algebra-for-radiation-exchange: mesubjects.net, Retrieved 2 June, 2019
- Understanding-classical-gray-body-radiation-theory: comsol.com, Retrieved 29 March, 2019
- S. Blundell, k. Blundell (2006). Concepts in thermal physics. Oxford university press. P. 247. Isbn 978-0-19-856769-1
- Grey-bodies-heat-exchange, radiation, heat-transfer: sbainvent.com, Retrieved 22 May, 2019

3

Understanding Convection

The heat transfer between particles due to the bulk movement of molecules within fluids such as gases and liquids is called convection. Advection, natural convection and granular convection are some of the types and mechanisms of convection. All these diverse aspects of convection have been carefully analyzed in this chapter.

Convection is a process by which heat is transferred by movement of a heated fluid such as air or water.

Natural convection results from the tendency of most fluids to expand when heated—*i.e.,* to become less dense and to rise as a result of the increased buoyancy. Circulation caused by this effect accounts for the uniform heating of water in a kettle or air in a heated room: the heated molecules expand the space they move in through increased speed against one another, rise, and then cool and come closer together again, with increase in density and a resultant sinking.

Forced convection involves the transport of fluid by methods other than that resulting from variation of density with temperature. Movement of air by a fan or of water by a pump are examples of forced convection.

Atmospheric convection currents can be set up by local heating effects such as solar radiation (heating and rising) or contact with cold surface masses (cooling and sinking). Such convection currents primarily move vertically and account for many atmospheric phenomena, such as clouds and thunderstorms.

ADVECTION

Advection is defined as a transfer of some property by the atmosphere or ocean. This term can actually be very vague unless there is a description of what exactly could be transferred. Most commonly, you will hear terms like 'cold-air advection' after the passing of a cold front where cold air is transferred into your region (usually from the North, Northwest or Northeast) resulting in a drop in temperatures. On the contrary, warm-air advection refers to the passing of a warm front that causes a transfer of warm air into your region (typically from the South, Southwest or Southeast) and increase temperatures.

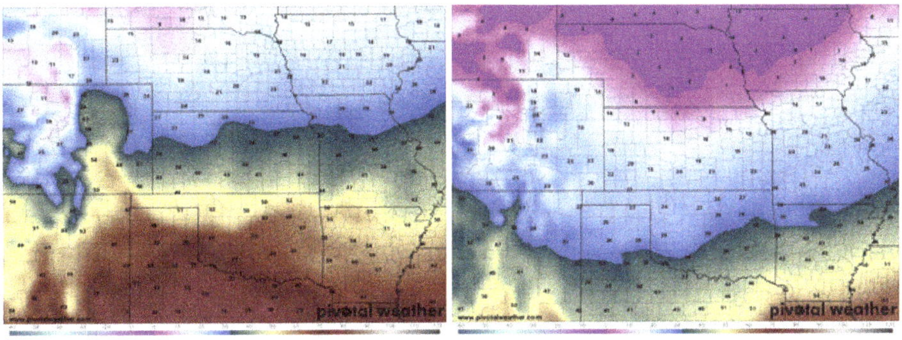

Hour temperature changes depicted from the GFS model on the night of March 1 to the morning of March 2 showing significant cold-air advection over 12 hours to the Oklahoma and Texas area.

While temperature is the most commonly paired term with advection, there are many more elements that can be transferred from one region to another. Wind can also aid in the transfer of more than just air temperature changes. Modification of air can refer to changing the humidity or dew point as well. Besides modifying the air with moisture, we can also see transfer of dirt, dust, salt or smoke.

The purpose of tracking changes to air temperature help to determine how localized elements will change during a given time. For example, transfer of moist, humid and unstable air to a different region that leads to convective storms (severe thunderstorms and tornados) is essential to track the potential for severe weather. Timing of potential impacts is essential to residents planning a commute or maintaining their personal safety.

Advection is a term that refers specifically to changes based on atmospheric or oceanic transfer. These are a few examples of what the term advection does not refer to: plowing snow from one area to another, watering your lawn, or salting an icy sidewalk or street.

If there are calm conditions at a station, that also does not mean that temperature will not change. Radiative cooling can occur at night with no solar heating and under dry conditions, which allows for heat to radiate from the surface that results in a temperature drop. Solar heating can cause an increase in temperature during the day merely from the warmth of the sun and no interaction from wind.

CONVECTION ZONE

A convection zone, convective zone or convective region of a star is a layer which is unstable to convection. Energy is primarily or partially transported by convection in such a region. In a radiation zone, energy is transported by radiation and conduction.

Stellar convection consists of mass movement of plasma within the star which usually forms a circular convection current with the heated plasma ascending and the cooled plasma descending.

The Schwarzschild criterion expresses the conditions under which a region of a star is unstable to convection. A parcel of gas that rises slightly will find itself in an environment of lower pressure than the one it came from. As a result, the parcel will expand and cool. If the rising parcel cools to a lower temperature than its new surroundings, so that it has a higher density than the surrounding gas, then its lack of buoyancy will cause it to sink back to where it came from. However, if the temperature gradient is steep enough (i. e. the temperature changes rapidly with distance from the center of the star), or if the gas has a very high heat capacity (i. e. its temperature changes relatively slowly as it expands) then the rising parcel of gas will remain warmer and less dense than its new surroundings even after expanding and cooling. Its buoyancy will then cause it to continue to rise. The region of the star in which this happens is the convection zone.

Main Sequence Stars

In main sequence stars more than 1.3 times the mass of the Sun, the high core temperature causes nuclear fusion of hydrogen into helium to occur predominantly via the carbon-nitrogen-oxygen (CNO) cycle instead of the less temperature-sensitive proton-proton chain. The high temperature gradient in the core region forms a convection zone that slowly mixes the hydrogen fuel with the helium product. The core convection zone of these stars is overlaid by a radiation zone that is in thermal equilibrium and undergoes little or no mixing. In the most massive stars, the convection zone may reach all the way from the core to the surface.

In main sequence stars of less than about 1.3 solar masses, the outer envelope of the star contains a region where partial ionization of hydrogen and helium raises the heat capacity. The relatively low temperature in this region simultaneously causes the opacity due to heavier elements to be high enough to produce a steep temperature gradient. This combination of circumstances produces an outer convection zone, the top of which is visible in the Sun as solar granulation. Low mass main sequences of stars, such as red dwarfs below 0.35 solar masses, as well as pre-main sequence stars on the Hayashi track, are convective throughout and do not contain a radiation zone.

In main sequence stars similar to the Sun, which have a radiative core and convective envelope, the transition region between the convection zone and the radiation zone is called the tachocline.

Red Giants

In red giant stars, and particularly during the asymptotic giant branch phase, the surface convection zone varies in depth during the phases of shell burning. This causes dredge-up events, short-lived very deep convection zones that transport fusion products to the surface of the star.

NATURAL CONVECTION

Natural convection is a type of flow, of motion of a liquid such as water or a gas such as air, in which the fluid motion is not generated by any external source (like a pump, fan, suction device, etc.) but by some parts of the fluid being heavier than other parts. The driving force for natural convection is gravity. For example if there is a layer of cold dense air on top of hotter less dense air, gravity pulls more strongly on the denser layer on top, so it falls while the hotter less dense air rises to take its place. This creates circulating flow: convection. As it relies of gravity, there is no convection in free-fall (inertial) environments, such as that of the orbiting International Space Station. Natural convection can occur when there are hot and cold regions of either air or water, because both water and air become less dense as they are heated. But, for example, in the world's oceans it also occurs due to salt water being heavier than fresh water, so a layer of salt water on top of a layer of fresher water will also cause convection.

Natural convection has attracted a great deal of attention from researchers because of its presence both in nature and engineering applications. In nature, convection cells formed from air raising above sunlight-warmed land or water are a major feature of all weather systems. Convection is also seen in the rising plume of hot air from fire, plate tectonics, oceanic currents (thermohaline circulation) and sea-wind formation (where upward convection is also modified by Coriolis forces). In engineering applications, convection is commonly visualized in the formation of microstructures during the cooling of molten metals, and fluid flows around shrouded heat-dissipation fins, and solar ponds. A very common industrial application of natural convection is free air cooling without the aid of fans: this can happen on small scales (computer chips) to large scale process equipment.

Parameters

Onset

The onset of natural convection is determined by the Rayleigh number (Ra). This dimensionless number is given by,

$$\mathbf{Ra} = \frac{\Delta\rho g L^3}{D\mu}$$

where,

- $\Delta\rho$ is the difference in density between the two parcels of material that are mixing.
- g is the local gravitational acceleration.
- L is the characteristic length-scale of convection: the depth of the boiling pot, for example.
- D is the diffusivity of the characteristic that is causing the convection.
- μ is the dynamic viscosity.

Natural convection will be more likely and/or more rapid with a greater variation in density between the two fluids, a larger acceleration due to gravity that drives the convection, and/or a larger distance through the convecting medium. Convection will be less likely and/or less rapid with more rapid diffusion (thereby diffusing away the gradient that is causing the convection) and/or a more viscous (sticky) fluid.

For thermal convection due to heating from below, as described in the boiling pot above, the equation is modified for thermal expansion and thermal diffusivity. Density variations due to thermal expansion are given by:

$$\Delta\rho = \rho_0 \beta \Delta T$$

where,

- ρ_0 is the reference density, typically picked to be the average density of the medium.
- β is the coefficient of thermal expansion.
- ΔT is the temperature difference across the medium.

The general diffusivity, D, is redefined as a thermal diffusivity, α.

$$D = \alpha$$

Inserting these substitutions produces a Rayleigh number that can be used to predict thermal convection.

$$\mathbf{Ra} = \frac{\rho_0 g \beta \Delta T L^3}{\alpha \mu}$$

Turbulence

The tendency of a particular naturally convective system towards turbulence relies on the Grashof number (Gr).

$$Gr = \frac{g\beta\Delta TL^3}{\nu^2}$$

In very sticky, viscous fluids (large ν), fluid movement is restricted, and natural convection will be non-turbulent.

Following the treatment of the previous subsection, the typical fluid velocity is of the order of $g\Delta\rho L^2 / \mu$, , up to a numerical factor depending on the geometry of the system. Therefore, Grashof number can be thought of as Reynolds number with the velocity of natural convection replacing the velocity in Reynolds number's formula. However In practice, when referring to the Reynolds number, it is understood that one is considering forced convection, and the velocity is taken as the velocity dictated by external constraints.

Behavior

The Grashof number can be formulated for natural convection occurring due to a concentration gradient, sometimes termed thermo-solutal convection. In this case, a concentration of hot fluid diffuses into a cold fluid, in much the same way that ink poured into a container of water diffuses to dye the entire space. Then:

$$Gr = \frac{g\beta\Delta CL^3}{\nu^2}$$

Natural convection is highly dependent on the geometry of the hot surface, various correlations exist in order to determine the heat transfer coefficient. A general correlation that applies for a variety of geometries is,

$$Nu = \left[Nu_0^{\frac{1}{2}} + Ra^{\frac{1}{6}} \left(\frac{f_4(Pr)}{300} \right)^{\frac{1}{6}} \right]^2$$

The value of f_4(Pr) is calculated using the following formula,

$$f_4(Pr) = \left[1 + \left(\frac{0.5}{Pr} \right)^{\frac{9}{16}} \right]^{\frac{-16}{9}}$$

Nu is the Nusselt number and the values of Nu_0 and the characteristic length used to calculate Ra are listed below:

Geometry	Characteristic length	Nu_0
Inclined plane	x (Distance along plane)	0.68

Inclined disk	9D/11 (D = diameter)	0.56
Vertical cylinder	x (height of cylinder)	0.68
Cone	4x/5 (x = distance along sloping surface)	0.54
Horizontal cylinder	(D = diameter of cylinder)	0.36

Warning: The values indicated for the Horizontal cylinder are wrong.

Natural Convection from a Vertical Plate

In this system heat is transferred from a vertical plate to a fluid moving parallel to it by natural convection. This will occur in any system wherein the density of the moving fluid varies with position. These phenomena will only be of significance when the moving fluid is minimally affected by forced convection.

When considering the flow of fluid is a result of heating, the following correlations can be used, assuming the fluid is an ideal diatomic, has adjacent to a vertical plate at constant temperature and the flow of the fluid is completely laminar.

$$Nu_m = 0.478(Gr^{0.25})$$

$$\text{Mean Nusselt Number} = Nu_m = h_m L/k$$

where,

- h_m = mean coefficient applicable between the lower edge of the plate and any point in a distance L (W/m². K).
- L = height of the vertical surface (m).
- k = thermal conductivity (W/m. K).

Grashof number = Gr = $[gL^3(t_s - t_\infty)]/v^2T$

where,

- g = gravitational acceleration (m/s²).
- L = distance above the lower edge (m)
- t_s = temperature of the wall (K).
- t_∞ = fluid temperature outside the thermal boundary layer (K).
- v = kinematic viscosity of the fluid (m²/s).
- T = absolute temperature (K).

When the flow is turbulent different correlations involving the Rayleigh Number (a function of both the Grashof Number and the Prandtl Number) must be used.

Note that the above equation differs from the usual expression for Grashof number because the value β has been replaced by its approximation $1/T$, which applies for ideal gases only (a reasonable approximation for air at ambient pressure).

Pattern Formation

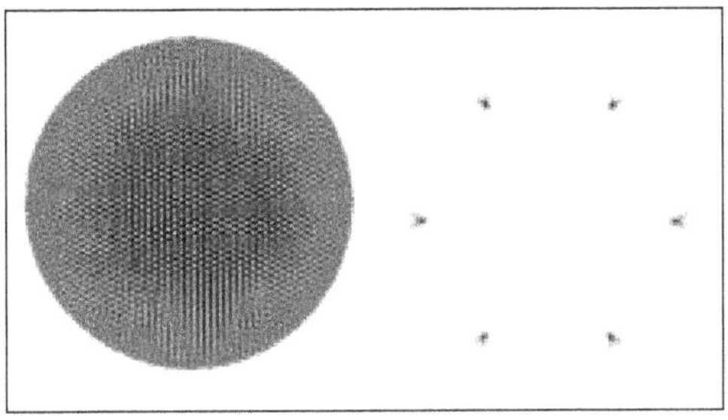

A fluid under Rayleigh-Bénard convection: the left picture represents the thermal field and the right picture its two-dimensional Fourier transform.

Convection, especially Rayleigh-Bénard convection, where the convecting fluid is contained by two rigid horizontal plates, is a convenient example of a pattern forming system.

When heat is fed into the system from one direction (usually below), at small values it merely diffuses (*conducts*) from below upward, without causing fluid flow. As the heat flow is increased, above a critical value of the Rayleigh number, the system undergoes a bifurcation from the stable *conducting* state to the *convecting* state, where bulk motion of the fluid due to heat begins. If fluid parameters other than density do not depend significantly on temperature, the flow profile is symmetric, with the same volume of fluid rising as falling. This is known as Boussinesq convection.

As the temperature difference between the top and bottom of the fluid becomes higher, significant differences in fluid parameters other than density may develop in the fluid due to temperature. An example of such a parameter is viscosity, which may begin to significantly vary horizontally across layers of fluid. This breaks the symmetry of the system, and generally changes the pattern of up- and down-moving fluid from stripes to hexagons, as seen at right. Such hexagons are one example of a convection cell.

As the Rayleigh number is increased even further above the value where convection cells first appear, the system may undergo other bifurcations, and other more complex patterns, such as spirals, may begin to appear.

Water Convection at Freezing Temperatures

Water is a fluid that does not obey the Boussinesq approximation. This is because its density varies nonlinearly with temperature, which causes its thermal expansion coefficient to be inconsistent near freezing temperatures. The density of water reaches a maximum at 4 °C and decreases as the temperature deviates. This phenomenon is investigated by experiment and numerical methods. Water is initially stagnant at 10 °C within a square cavity. It is differentially heated between the two vertical walls, where the left and right walls are held at 10 °C and 0 °C, respectively. The density anomaly manifests in its flow pattern. As the water is cooled at the right wall, the density increases, which accelerates the flow downward. As the flow develops and the water cools further, the decrease in density causes a recirculation current at the bottom right corner of the cavity.

Another case of this phenomenon is the event of super-cooling, where the water is cooled to below freezing temperatures but does not immediately begin to freeze. Under the same conditions as before, the flow is developed. Afterward, the temperature of the right wall is decreased to −10 °C. This causes the water at that wall to become supercooled, create a counter-clockwise flow, and initially overpower the warm current. This plume is caused by a delay in the nucleation of the ice. Once ice begins to form, the flow returns to a similar pattern as before and the solidification propagates gradually until the flow is redeveloped.

Mantle Convection

Convection within Earth's mantle is the driving force for plate tectonics. Mantle convection is the result of a thermal gradient: the lower mantle is hotter than the upper mantle, and is therefore less dense. This sets up two primary types of instabilities. In the first type, plumes rise from the lower mantle, and corresponding unstable regions of lithosphere drip back into the mantle. In the second type, subducting oceanic plates (which largely constitute the upper thermal boundary layer of the mantle) plunge back into the mantle and move downwards towards the core-mantle boundary. Mantle convection occurs at rates of centimeters per year, and it takes on the order of hundreds of millions of years to complete a cycle of convection.

Neutrino flux measurements from the Earth's core show the source of about two-thirds of the heat in the inner core is the radioactive decay of ^{40}K, uranium and thorium. This has allowed plate tectonics on Earth to continue far longer than it would have if it were simply driven by heat left over from Earth's formation; or with heat produced from gravitational potential energy, as a result of physical rearrangement of denser portions of the Earth's interior toward the center of the planet (i.e., a type of prolonged falling and settling).

RAYLEIGH–BÉNARD CONVECTION

Bénard–Rayleigh convection is a type of natural convection, occurring in a plane horizontal layer of fluid heated from below, in which the fluid develops a regular pattern of convection cells known as Bénard cells. Bénard–Rayleigh convection is one of the most commonly studied convection phenomena because of its analytical and experimental accessibility. The convection patterns are the most carefully examined example of self-organizing nonlinear systems.

Buoyancy, and hence gravity, are responsible for the appearance of convection cells. The initial movement is the upwelling of lesser density fluid from the heated bottom layer. This upwelling spontaneously organizes into a regular pattern of cells.

Physical Processes

The features of Bénard convection can be obtained by a simple experiment first conducted by Henri Bénard, a French physicist, in 1900.

Development of Convection

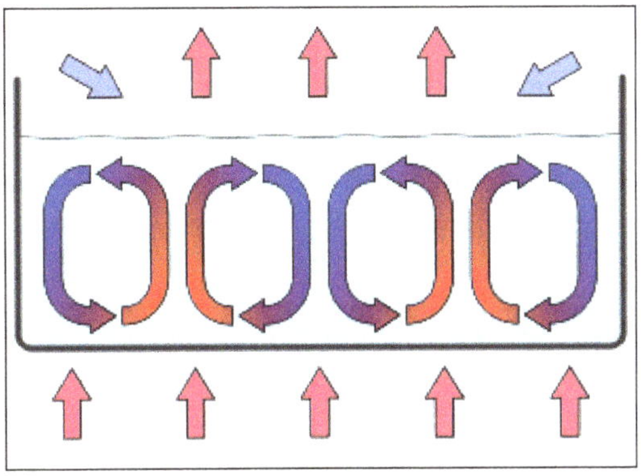

Convection cells in a gravity field.

The experimental set-up uses a layer of liquid, e.g. water, between two parallel planes. The height of the layer is small compared to the horizontal dimension. At first, the temperature of the bottom plane is the same as the top plane. The liquid will then tend towards an equilibrium, where its temperature is the same as its surroundings. (Once there, the liquid is perfectly uniform: to an observer it would appear the same from any position. This equilibrium is also asymptotically stable: after a local, temporary perturbation of the outside temperature, it will go back to its uniform state, in line with the second law of thermodynamics).

Then, the temperature of the bottom plane is increased slightly yielding a flow of thermal energy conducted through the liquid. The system will begin to have a structure of thermal conductivity: the temperature, and the density and pressure with it, will vary linearly between the bottom and top plane. A uniform linear gradient of temperature will be established. (This system may be modelled by statistical mechanics).

Once conduction is established, the microscopic random movement *spontaneously* becomes ordered on a macroscopic level, forming Benard convection cells, with a characteristic correlation length.

Convection Features

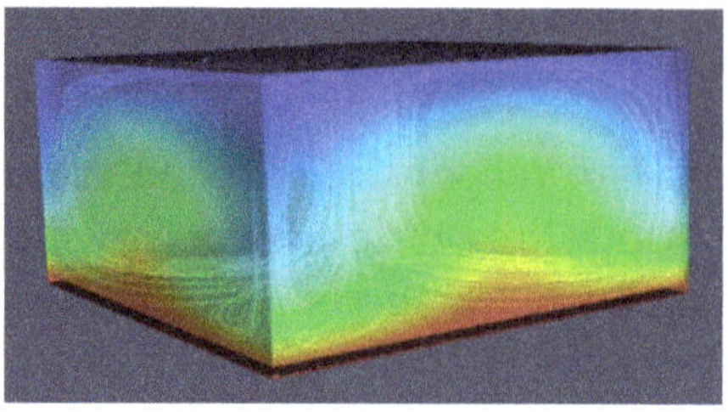

Simulation of Rayleigh–Bénard convection in 3D.

The rotation of the cells is stable and will alternate from clock-wise to counter-clock-wise horizontally; this is an example of spontaneous symmetry breaking. Bénard cells are metastable. This means that a small perturbation will not be able to change the rotation of the cells, but a larger one could affect the rotation; they exhibit a form of hysteresis.

Moreover, the deterministic law at the microscopic level produces a non-deterministic arrangement of the cells: if the experiment is repeated, a particular position in the experiment will be in a clockwise cell in some cases, and a counter-clockwise cell in others. Microscopic perturbations of the initial conditions are enough to produce a non-deterministic macroscopic effect. That is, in principle, there is no way to calculate the macroscopic effect of a microscopic perturbation. This inability to predict long-range conditions and sensitivity to initial-conditions are characteristics of chaotic or complex systems (i.e., the butterfly effect).

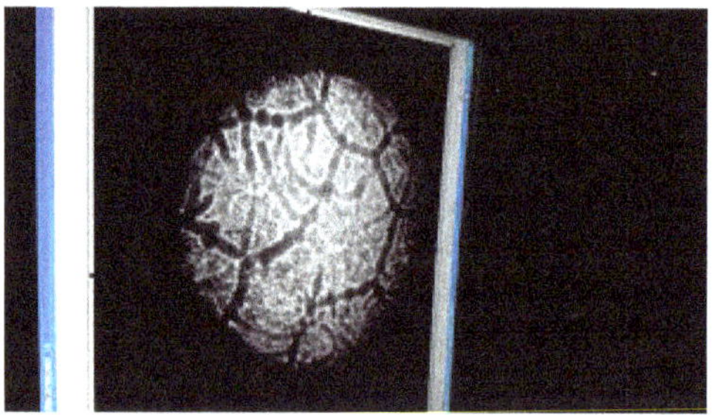

Turbulent Rayleigh–Bénard convection.

If the temperature of the bottom plane was to be further increased, the structure would become more complex in space and time; the turbulent flowwould become chaotic.

Convective Bénard cells tend to approximate regular right hexagonal prisms, particularly in the absence of turbulence, although certain experimental conditions can result in the formation of regular right square prisms or spirals.

The convective Bénard cells are not unique and will usually appear only in the surface tension driven convection. In general the solutions to the Rayleigh and Pearson analysis (linear theory) assuming an infinite horizontal layer gives rise to degeneracy meaning that many patterns may be obtained by the system. Assuming uniform temperature at the top and bottom plates, when a realistic system is used (a layer with horizontal boundaries) the shape of the boundaries will mandate the pattern. More often than not the convection will appear as rolls or a superposition of them.

Rayleigh–Bénard Instability

Since there is a density gradient between the top and the bottom plate, gravity acts trying to pull the cooler, denser liquid from the top to the bottom. This gravitational force is opposed by the viscous damping force in the fluid. The balance of these two forces is expressed by a non-dimensional parameter called the Rayleigh number. The Rayleigh number is defined as:

$$Ra_L = \frac{g\beta}{\nu\alpha}(T_b - T_u)L^3$$

where,

- T_u is the temperature of the top plate.
- T_b is the temperature of the bottom plate.
- L is the height of the container.
- g is the acceleration due to gravity.
- ν is the kinematic viscosity.
- α is the Thermal diffusivity.
- β is the Thermal expansion coefficient.

As the Rayleigh number increases, the gravitational forces become more dominant. At a critical Rayleigh number of 1708, instability sets in and convection cells appear.

The critical Rayleigh number can be obtained analytically for a number of different boundary conditions by doing a perturbation analysis on the linearized equations in the stable state. The simplest case is that of two free boundaries, which Lord Rayleigh solved in 1916, obtaining $Ra = \frac{27}{4}\pi^4 \approx 657.51$. In the case of a rigid boundary at the bottom and a free boundary at the top (as in the case of a kettle without a lid), the critical Rayleigh number comes out as Ra = 1,100.65.

Effects of Surface Tension

In case of a free liquid surface in contact with air, buoyancy and surface tension effects will also play a role in how the convection patterns develop. Liquids flow from places of lower surface tension to places of higher surface tension. This is called the Marangoni effect. When applying heat from below, the temperature at the top layer will show temperature fluctuations. With increasing temperature, surface tension decreases. Thus a lateral flow of liquid at the surface will take place, from warmer areas to cooler areas. In order to preserve a horizontal (or nearly horizontal) liquid surface, cooler surface liquid will descend. This down-welling of cooler liquid contributes to the driving force of the convection cells. The specific case of temperature gradient-driven surface tension variations is known as thermo-capillary convection, or Bénard–Marangoni convection.

GRANULAR CONVECTION

Granular convection, or granular segregation, is a phenomenon where granular material subjected to shaking or vibration will exhibit circulation patterns similar to types of fluid convection. It is sometimes described as the Brazil nut effect when the largest particles end up on the surface of a granular material containing a mixture of variously sized objects; this derives from the example of a typical container of mixed nuts, where the largest will be Brazil nuts. The phenomenon is also known as the muesli effect since it is seen in packets of breakfast cereal containing particles of different sizes but similar density, such as muesli mix.

Under experimental conditions, granular convection of variously sized particles has been observed forming convection cells similar to fluid motion. The convection of granular flows is becoming a well-understood phenomenon.

Explanation

It may be counterintuitive to find that the largest and (presumably) heaviest particles rise to the top, but several explanations are possible:

- The center of mass of the whole system (containing the mixed nuts) in an arbitrary state is not optimally low; it has the tendency to be higher due to there being more empty space around the larger Brazil nuts than around smaller nuts. When the nuts are shaken, the system has the tendency to move to a lower energy state, which means moving the center of mass down by moving the smaller nuts down and thereby the Brazil nuts up.
- Including the effects of air in spaces between particles, larger particles may become buoyant or sink. Smaller particles can fall into the spaces underneath a larger particle after each shake. Over time, the larger particle rises in the mixture. (According to Heinrich Jaeger, "[this] explanation for size separation might work in situations in which there is no granular convection, for example for containers with completely frictionless side walls or deep below the surface of tall containers (where convection is strongly suppressed). On

the other hand, when friction with the side walls or other mechanisms set up a convection roll pattern inside the vibrated container, we found that the convective motion immediately takes over as the dominant mechanism for size separation").

- The same explanation without buoyancy or center of mass arguments: As a larger particle moves upward, any motion of smaller particles into the spaces underneath blocks the larger particle from settling back in its previous position. Repetitive motion results in more smaller particles slipping beneath larger particles. A greater density of the larger particles has no effect on this process. Shaking is not necessary; any process which raises particles and then lets them settle would have this effect. The process of raising the particles imparts potential energy into the system. The result of all the particles settling in a different order may be an increase in the potential energy—a raising of the center of mass.
- When shaken, the particles move in vibration-induced convection flow; individual particles move up through the middle, across the surface, and down the sides. If a large particle is involved, it will be moved up to the top by convection flow. Once at the top, the large particle will stay there because the convection currents are too narrow to sweep it down along the wall.

The phenomenon is related to Parrondo's paradox in as much as the Brazil nuts move to the top of the mixed nuts against the gravitational gradient when subjected to random shaking.

Granular convection has been probed by the use of MRI where convection rolls similar to those in fluids (Bénard cells) can be visualized.

Applications

Manufacturing

The effect is of serious interest for some manufacturing operations; once a homogeneous mixture of granular materials has been produced, it is usually undesirable for the different particle types to segregate. Several factors determine the severity of the Brazil nut effect, including the sizes and densities of the particles, the pressure of any gas between the particles, and the shape of the container. A rectangular box (such as a box of breakfast cereal) or cylinder (such as a can of nuts) works well to favour the effect, while a cone-shaped container results in what is known as the reverse Brazil nut effect.

Astronomy

In astronomy, it is also seen in some low density, or rubble pile asteroids, for example the asteroid 25143 Itokawa.

Geology

In geology, the effect is common in formerly glaciated areas such as New England and areas in regions of permafrost where the landscape is shaped into hummocks by frost heave — new stones appear in the fields every year from deeper underground. Horace Greeley noted "Picking stones is a never-ending labor on one of those New England farms. Pick as closely as you may, the next plowing turns up a fresh eruption of boulders and pebbles, from the size of a hickory nut to that of a tea-kettle." A hint to the cause appears in his further description that "this work is mainly to be done in March or April, when the earth is saturated with ice-cold water". Underground water freezes, lifting all particles above it. As the water starts to melt, smaller particles can settle into the opening spaces while larger particles are still raised. By the time ice no longer supports the larger rocks, they are at least partially supported by the smaller particles that slipped below them. Repeated freeze-thaw cycles in a single year speeds up the process.

This phenomenon is one of the causes of inverse grading which can be observed in many situations including soil liquefaction during earthquakes or mudslides. Granular convection is also exemplified by debris flow, which is a fast moving, liquefied landslide of unconsolidated, saturated debris that looks like flowing concrete. These flows can carry material ranging in size from clay to boulders, including woody debris such as logs and tree stumps. Flows can be triggered by intense rainfall, glacial melt, or a combination of the two.

FORCED CONVECTION

Forced convection is a special type of heat transfer in which fluids are forced to move, in order to increase the heat transfer. This forcing can be done with a ceiling fan, a pump, suction device, or other.

Convection is a heat transfer mechanism where heat moves from one place to another through fluid currents. Forced convection is simply using this mechanism in a useful way to heat or cool a home efficiently, such as using a fan.

Many people are familiar with the statement that "heat rises". This is a simplification of the idea that hot fluids are almost always less dense than the same fluid when cold, but there are exceptions. This difference in density makes hotter material naturally end up on top of cooler material due to the higher buoyancy of the hotter material.

Natural convection can create a noticeable difference in temperature within a home. Often this becomes places where certain parts of the house are warmer and certain parts are cooler. Forced convection creates a more uniform and therefore comfortable temperature throughout the entire home. This reduces cold spots in the house, reducing the need to crank the thermostat to a higher temperature, or putting on sweaters.

Operation

A floor heat register is part of the HVAC system that creates forced convection in a home.

Creating forced convection is as easy as turning on a fan. Air is heated in the furnace and pushed through the house by the *blower*, which is a fan inside the ventilation system. This blower outputs a specific quantity of air, and this output air flow is divided among all of the output grills (also called heater vents) in a home. Once it has traveled through the vents by being pushed through by fans, the warm, treated air is ejected through floor or ceiling vents into the rooms of a house. With help from natural convection this air then travels through the room, warming the room as it rises to the top by natural convection and slowly falls down to the floor as it cools. The system of heating the air and pushing it throughout the house to warm it then begins again.

How the treated air gets to the output vents makes a difference, as the structure of the ductwork can create resistance to airflow at elbows, divisions, or places where the ductwork size changes. This change in turn effects how well this forced air system can heat a home since they are all sharing the output flow of air from one source —the furnace. Therefore, properly planning out the ducting is important. As a general rule, the best way for air to move through a duct is to have a straight duct, that is round in shape with a smooth inside wall—since curves and corners resist air flow. Wherever possible, this guideline should be followed to ensure that the air that is being forced out by the furnace heats the house properly. In addition, ensuring output vents are not covered by furniture or installed behind curtains ensures that the warm air output by the furnace is able to circulate throughout the room.

It is a common misconception that the more air that flows from a fan—or the more a fan "pushes" the air— the greater the effects of forced convection will occur, due to the large amount of heated or cooled air being pushed out by the fan. However, this is not entirely true. Part of how air moves through a home or other building has to do with the pressureand temperature that exists in the room before more air is pushed through. For example, if a room has a cold spot and the goal is to heat the room evenly, the pressure change in the area between the cold and warm spots, known as transitional "warm" area, factors into how well a fan will be able to move warm air to the cold area. If the pressure drop in this warm area is higher, there will be a lower flow rate of air into the cold section of the room, making it more difficult for the fan to push warm air into this section. This phenomenon is known as the *pressure drop* over the *heat sink* and it can be summed up easily by saying that it is more difficult for a fan to push warm or cool air through a region between two areas of different temperatures that also has a large pressure difference across its boundary.

Mechanism of Forced Convection

Convection is a complex heat transfer method, but can be expressed by Newton's Law of Heating and Cooling:

$$q^*_{conv} = h(T_s - T_\infty)$$

Which simply says that the rate of convection heat transfer (q^*_{conv}), expressed in the units (W/m^2) is proportional to the difference between the initial temperature of the material (T_s) and the final temperature of the material (T_∞) through a proportionality constant *h*. The rate of heat transfer is also strongly dependent on the roughness and shape of the material being heated. Newton's Law of Heating and Cooling changes depending on whether or not the convection is forced. For natural cooling, the *h* value is equal to a certain number. However, by forcing convection and pushing heated or cooled air from one place to another one is able to *change* this proportionality constant and heat or cool an object more quickly.

Ceiling Fans

In the winter, ceiling fans should rotate clockwise to pull cool air up from the room and force warm air downwards, creating an updraft.

The use of ceiling fans in a home also represents a different type of forced convection. Ceiling fans can be used in both the winter and the summer, but their settings must be different in order to perform the desired task. In the summer months, the fan is generally set to a higher speed. The angle of the blades forces air down through the room. Generally this corresponds to a counter-clockwise rotation when looking at the fan from below. This downward breeze aids in the evaporation of perspiration from inhabitants of the home, cooling them. In the winter months, the fan must be used on a slower speed. The blades also spin in a different direction, generally clockwise when looking from below the fan, which pulls the cooler air up from the lower parts of the room. The cooler air from below then mixes with the warmer air that has risen and mixes the two, distributing warmer air throughout the building.

In the summer, ceiling fans should rotate counterclockwise
to mix warm air and force a cool breeze downwards, creating a downdraft.

References

- Convection, science: britannica.com, Retrieved 23 February, 2019
- Behrend, r.; maeder, a. (2001). "formation of massive stars by growing accretion rate". Astronomy and astrophysics. 373: 190. Arxiv:astro-ph/0105054. Bibcode: 2001a&a...373..190b. Doi:10.1051/0004-6361:20010585
- Donald l. Turcotte; gerald schubert. (2002). Geodynamics. Cambridge: cambridge university press. Isbn 978-0-521-66624-4
- Moore, emily b.; molinero, valeria (november 2011). "structural transformation in supercooled water controls the crystallization rate of ice". Nature. 479 (7374): 506–508. Arxiv:1107.1622. Doi:10.1038/nature10586. Issn 0028-0836. Pmid 22113691
- abbott, derek (2009). "developments in parrondo›s paradox". Applications of nonlinear dynamics. Springer. Pp. 307–321. Isbn 978-3-540-85631-3
- Forced-convection, encyclopedia: energyeducation.ca, Retrieved 2 February, 2019

4

Condensation and Boiling Heat Transfer

Condensation heat transfer is the heat transfer accompanied by condensation. The heat transfer accompanied by boiling is referred to as boiling heat transfer. The different phenomena related to conduction and boiling heat transfer are film condensation, nucleate boiling, etc. These topics elaborated in this chapter will help in gaining a better perspective about condensation and boiling heat transfer.

CONDENSATION HEAT TRANSFER

Condensation heat transfer plays an important role in many engineering applications, notably electric power generation, process industries, refrigeration and air-conditioning. Many different physical phenomena are involved in the condensation process, their relative importance depending on the circumstances and application.

When a liquid and its vapor are in contact, molecules pass from liquid to vapor and from vapor to liquid. Condensation occurs when the number of molecules entering the liquid phase exceeds that of leaving molecules. Under these circumstances, the temperature of the vapor in the immediate vicinity (a few mean free paths) of the vapor-liquid interface is higher than that of the liquid.

The interface temperature dropincreases with increasing condensation rate and with decreasing pressure but in most circumstances (an exception being the case of liquid metals), this is very small and equilibrium conditions at the interface can be assumed. A brief summary of interface matter transfer during condensation is given by Niknejad and Rose.

The most common and best understood case of condensation heat transfer is that of film *condensation* of a pure quiescent vapor on a solid surface. The problem of calculating the heat transfer rate for a plane vertical surface and for a horizontal cylinder with uniform surface temperatures, and where the condensate flow is laminar and governed only by gravity and viscous forces, has been solved by Nusselt. Nusselt's main assumptions are that heat transfer across the condensate film is by pure conduction, the effect of vapor drag in supporting the falling condensate film is negligible, and the properties of the condensate may be taken to be uniform across the film, i.e., are essentially independent of temperature.

The well-known Nusselt equations are:

For the vertical plane surface, the mean value of Nusselt number is,

$$Nu = 0.943\left\{\frac{\rho\Delta\rho g h_{lg} L^3}{\eta\lambda\Delta T}\right\}^{1/4},$$

and for the horizontal tube,

$$Nu = 0.728\left\{\frac{\rho\Delta\rho g h_{lg} d^3}{\eta\lambda\Delta T}\right\}^{1/4},$$

where ρ is the liquid density, g is acceleration due to gravity, h_{lg} is the latent heat of evaporation, η is the liquid viscosity, λ is the liquid thermal conductivity, $\Delta\rho$ is the density difference between vapor and condensate, ΔT is the vapor-to-surface temperature difference; L is the plate height; and d, the tube diameter. For the case of the tube, the additional assumption that film thickness is small compared with the tube radius is needed. Since the theory predicts that the radial film thickness tends to infinity at the bottom of the tube, this assumption is evidently invalid for the lower part of the tube. The fact that the heat transfer rate is inaccurate where the film becomes thick is relatively unimportant because it is small and makes a minor contribution to the total heat transfer rate for the tube. $Nu = 0.943\left\{\frac{\rho\Delta\rho g h_{lg} L^3}{\eta\lambda\Delta T}\right\}^{1/4}$ and $Nu = 0.728\left\{\frac{\rho\Delta\rho g h_{lg} d^3}{\eta\lambda\Delta T}\right\}^{1/4}$, have been well verified experimentally for condensation of pure (only one molecular constituent) vapor. In order to obtain the constant in $Nu = 0.728\left\{\frac{\rho\Delta\rho g h_{lg} d^3}{\eta\lambda\Delta T}\right\}^{1/4}$, numerical integration is required twice. Nusselt's slightly inaccurate value ($0.8024\,(2/3)^{1/4} = 0.725$) is due to his use of planimetry.

As discussed by Rose, more recent theoretical studies in which the effects of inertia and convection in the condensate film and vapor drag on the condensate surface are included have shown that these effects are unimportant. More recently, it has been shown by Memory and Rose that the effect of variable wall temperature, which occurs in practice during condensation on a horizontal tube, also has a negligible effect on the *mean* heat transfer rate. Thus, $Nu = 0.943\left\{\frac{\rho\Delta\rho g h_{lg} L^3}{\eta\lambda\Delta T}\right\}^{1/4}$ and $Nu = 0.728\left\{\frac{\rho\Delta\rho g h_{lg} d^3}{\eta\lambda\Delta T}\right\}^{1/4}$, can be used with confidence for condensation of pure "stationary" vapors when the condensate flow is laminar.

In the case of significant vapor flow rate over the condensing surface, the effect of drag on the condensate becomes significant. In some circumstances, the effect of vapor drag overwhelms that of gravity. Since 1960, vapor shear stress effects have been studied extensively. Some of the more important contributions are described by Rose.

The relative importance of vapor shear stress and gravity on the motion of the condensate film is measured by the dimensionless parameter $F=\eta h_{lg} gx / \lambda \Delta T u^2_{\infty}$, where x is the relevant linear dimension (plate height or tube diameter) and u_{∞} is the free-stream vapor velocity. For downward vapor flow over a horizontal tube, an approximate analysis gives.

$$\mathrm{Nu_d}\, \overline{\mathrm{Re}}_d^{-1/2} \frac{0.9+0.728F^{1/2}}{\left(1+3.44F^{1/2}+F\right)^{1/4}}$$

$\mathrm{Nu_d}$ is the mean Nusselt number for the condensate film and $\overline{\mathrm{Re}}_d$ is a Reynolds number using the vapor approach velocity and condensate properties. $\mathrm{Nu_d}\, \overline{\mathrm{Re}}_d^{-1/2} \frac{0.9+0.728F^{1/2}}{\left(1+3.44F^{1/2}+F\right)^{1/4}}$ which indicates that for $F > 10$, gravity dominates while for $F < 0.1$, vapor shear stress is controlling—agrees quite well with experimental data from several investigations using various condensing fluids.

When the vapor contains more than one molecular species, the problem is complicated by diffusion of species in the vapor. For example, for a two-constitutent vapor where only one constituent condenses (e.g., steam-air), the mixture is rich in the *noncondensing gas* near the interface where vapor molecules are removed. The tendency is for noncondensing gas molecules to diffuse away from the interface so that in the steadystate, the rate at which gas molecules arrive at the surface with the condensing vapor is equal to their diffusion rate away from the surface. Even in the absence of forced convection of the vapor-gas mixture, the density difference, which results from the composition difference between that of the bulk vapor and that of the vapor adjacent to the interface, leads to natural convection. The process by which the steadystate is reached is therefore one of diffusion in the presence of convection. The fact that the vapor-gas mixture adjacent to the condensate surface is rich in non-condensing gas causes the temperature at the interface to be lower than in the bulk. Assuming equilibrium at the interface, the temperature is equal to the saturation temperature corresponding to the partial pressure of the vapor, which may be significantly lower than in the more remote vapor. Composition (or partial pressure) and temperature boundary layer are set up in the vapor adjacent to the interface. This gives an effective heat transfer resistance since the temperature drop across the condensate film, and hence the heat transfer rate, is significantly reduced. Detailed boundary layer solutions of this problem for free and forced convection, notably by Koh, Sparrow, Fujii et al., have been given. Earlier works are discussed and approximate equations are given by Rose for the free convection problem, and Rose for the forced convection case. When two or more constitutents of the vapor condense together, the situation is similar to that described above since the more volatile constituent is more concentrated at the condensate surface. Extensive treatments of these problems for the case of the plane vertical condensing surface and laminar flow of vapor and condensate have been given by Fujii.

Approximate methods, based on the *analogy* between heat and diffusive mass transfer in the vapor, are widely used for multiconstituent problems. The essence of the method is that the differential equations expressing conservation of energy and molecular species can be arranged in identical form by appropriate nondimensionalization. Known results (theoretical or experimental) for heat transfer problems are used to infer results for the corresponding mass transfer diffusion problems. The method has wide utility but is approximate since the boundary condition on the normal vapor velocity at the surface is not the same for the heat and mass transfer problems. For the heat transfer problem, the normal velocity at the (solid) heat-exchanger wall is zero. In the case of condensation, where molecules pass through the vapor-liquid interface, the normal velocity is not zero. The validity of the analogy depends on the smallness of the "suction parameter" $(-v_0/u)Re^{1/2}$, where v_0 is the normal outward (i.e., negative for condensation) vapor velocity at the condensate surface, u is the free-stream velocity parallel to the surface and Re is the free-stream vapor Reynolds number. The results are strictly correct only in the limiting case of zero condensation rate. A widely-used approximate *stagnant film model* extends the range of validity of the analogy. The heat-mass transfer analogy and the stagnant film model are discussed by Lee and Rose, Butterworth.

The foregoing refers exclusively to laminar flow conditions. For tall condensing surfaces, or under conditions of high vapor shear stress, transition to turbulent flow in the condensate film may occur. This brings to the problem unresolved difficulties associated with the general problem of Turbulence. For moderate or low vapor velocities, the "effective height" of the surface for a horizontal tube (≈d) is small and laminar flow of the condensate is expected. Moreover, since turbulent mixing enhances heat transfer across the condensate film, the Nusselt solution $Nu=0.728\left\{\frac{\rho\Delta\rho g h_{1g} d^3}{\eta\lambda\Delta T}\right\}^{1/4}$ is conservative and is widely used in design calculations. Various models for turbulent film condensation from an essentially stationary vapor exist in the literature and predict somewhat different results. For gravity-dominated flow, transition to turbulence has been found to occur at film Reynolds numbers, $4\Gamma/\eta$, rather lower than 2 000, where Γ is the condensate flow rate per width of surface. In the presence of high-vapor shear stress, the problem is more complicated. Turbulent film condensation on a vertical surface under free-convection conditions can, in principle, be analyzed by an approach similar to that used for single-phase pipe flow. However, this problem is relatively unimportant since in practice, condensing surfaces are usually not sufficiently tall for turbulence to occur under purely free-convection conditions. Significant shear stress, due to vapor flow along the condensate surface, promotes the onset of turbulence. The analysis is then more complicated, particularly when the vapor flow is not in the same direction of gravity.

Turbulence is more often encountered for condensation inside a tube. In this case, the problem is generally complicated by the presence of significant vapor shear stress on the condensate film since even when all of the vapor is condensed in the tube, the shear stress for a portion of the tube towards the inlet end is generally significant owing to the high vapor velocity resulting from the small tube cross-section. Condensation inside tubes is beset with all the problems and uncertainties of Two-Phase Flow. The only case which can be analyzed wholly satisfactorily is that of downward vapor flow in a vertical tube with a laminar condensate film on the wall (stratified flow). In this case, the problem is the same as that for external flow, except that account must be taken of the progressive reduction of vapor flow rate, and hence shear stress, due to condensation. Approaches used in other cases are outlined by Butterworth.

In many practical applications, condensation occurs in bundles or banks of horizontal tubes (shell-side condensation). In these cases, there is the additional complication of *inundation* (condensate from higher or upstream tubes falling or impinging on lower or downstream tubes). This leads to thicker condensate films on the inundated tubes. At the same time, the condensate film on inundated tubes is disturbed and the heat transfer coefficient may be enhanced. Nusselt's approach for a vertical, in-line column of horizontal tubes assumes that condensate drains to lower tubes in the form of a continuous laminar film. This leads to a simple expression for the average heat transfer coefficient for a column of N tubes:

$$\bar{\alpha}_N = \alpha_1 N^{-1/4}$$

Where α_1, for the uppermost tube, is given by $\mathrm{Nu} = 0.728\left\{\frac{\rho\Delta\rho g h_{1g} d^3}{\eta\lambda\Delta T}\right\}^{1/4}$, In view of the more probable mode of *drainage* or inundation with condensate film disturbance due to splashing from droplets, columns or unstable broken films of liquid, it is not surprising that $\bar{\alpha}_N = \alpha_1 N^{-1/4}$ has been found to be overconservative. Many experimental studies of condensation on tube banks have been made. The data are widely-scattered owing primarily to the effects of noncondensing gases, turbulence and vapor velocity. Various correlations have been proposed and approximate methods used in practiced are discussed by Butterworth. Numerous techniques for enhancement of film condensation heat transfer have been proposed.Notable amongst these, for shell-side condensation, are low (fin height small in relation to tube diameter) integral-fin tubes. In this case, it is found that for horizontal tubes, the enhancement of the heat transfer coefficient can significantly exceed the increase in surface area due to the presence of the fins. The reasons for this are: 1) the vertical or near-vertical fin flanks have small heights so that the heat transfer coefficients are large $Nu=0.943\left\{\frac{\rho\Delta\rho g h_{1g} L^3}{\eta\lambda\Delta T}\right\}^{1/4}$, $Nu=0.728\left\{\frac{\rho\Delta\rho g h_{1g} d^3}{\eta\lambda\Delta T}\right\}^{1/4}$, surface tension effects give rise to an additional mechanism for draining condensate from parts of the surface. The latter arises from the pressure gradient set up in the presence of a condensate surface of varying curvature, e.g., for the two-dimensional case:

$$\frac{dp}{ds}=\sigma\frac{d(r^{-1})}{ds}$$

where P is pressure, s is distance measured along the surface, σ is surface tension and r is the local radius of curvature of the condensate film. At the same time, surface tension has a detrimental effect on the heat transfer coefficient due to capillary retention of condensate between the fins and the consequent "blanketing" of heat transfer surface on the lower part of the surface. The extent of condensate retention can be calculated from:

$$\phi=\cos^{-1}\{(4\sigma\cos\beta / pgbd_0)-1\}$$

as formulated by Honda et al. In equation $\phi=\cos^{-1}\{(4\sigma\cos\beta / pgbd_0)-1\}$, φ is the angle measured from the top of the tube to the position where the interfin tube space is filled with retained condensate; β is the angle between the fin flank and radial plane; b is the distance between adjacent fins measured at the fin tip; and d_0 is the tube diameter over the fins.

In an early theoretical solution of the problem of condensation on low-finned tubes by Beatty and Katz, the vertical fin-flanks and horizontal interfin tube spaces were treated on the basis of the Nusselt theory and effects of surface tension were ignored. This simple approach proved quite successful in practice for relatively low-surface tension fluids. This is partly because condensate retention in this case is small, and partly because the beneficial and detrimental effects of surface tension tend, to some extent, to nullify each other. More detailed and complicated models have been proposed, notably by Honda and Nozu. These require numerical solution and are less readily applied than the simple analytical result of Beatty and Katz. The various models are discussed by Marto. A recent semi-empirical approach, which includes surface tension effects, has been given by Rose. The result, in the form of an equation for the "enhancement ratio", is in good agreement with experimental data from seven investigations using four condensing fluids and 41 tube/fin geometries.

The foregoing relates to the case when the condensate wets the condensing surface and forms a continuous film. When the surface is not wetted, a quite different mode, namely Dropwise Condensation, may occur. In this case, minute droplets form at nucleation sites on the surface and growth takes place by condensation and coalescence with neighbors until drops reach a size at which they are removed from the surface by gravity or vapor shear stress. Moving drops sweep up stationary drops in their path, making available new area for condensation. The maximum-to-minimum drop size is around 10^6. To date, dropwise condensation has only been obtained with a few high-surface tension fluids (notably water). A nonwetting agent or "promoter" is required to promote dropwise condensation on metal surfaces. Heat transfer coefficients for dropwise condensation are much higher than those for film condensation under the same conditions. For steam at atmospheric pressure, the factor is around 20. Not surprisingly, this has stimulated research work over the past 60 years with the aim of finding an effective promoter. Although good promoters (e.g., dioctadecyl disulfide) are available, which form stable monomolecular layers on copper or copper-containing surfaces, and give dropwise condensation for hundreds or thousands of hours under clean laboratory conditions, effective promoters for use under industrial conditions have yet to be found. Recent surveys of dropwise condensation heat transfer have been given by Tanasawa and Rose.

In some applications, such as desalination and geothermal power plant, use is made of direct contact condensation. This is a term applied to processes wherein vapor condenses on subcooled liquid drops, sprays, jets, films or in a liquid pool. Various types of equipment and the relevant mechanisms for direct contact condensation are given by Jacobs

FILM CONDENSATION

In film condensation, the condensate wets the surface and forms a liquid film on the surface that slides down under the influence of gravity. Film condensation results in low heat transfer rates as the film of condensate impedes the heat transfer. The thickness of the film formed depends on many parameters including orientation of the surface, viscosity, rate of condensation etc. The film increases a thermal resistance to heat flow between the surface and the vapour. The rate of heat transfer is reduced because of this resistance.

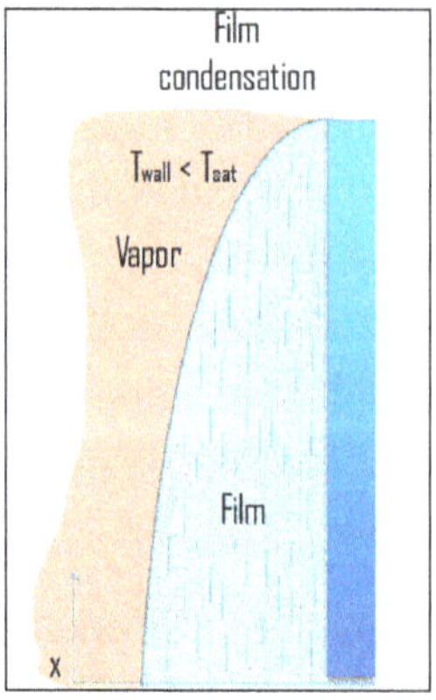

Condensation in Power Plants – Main Condensator

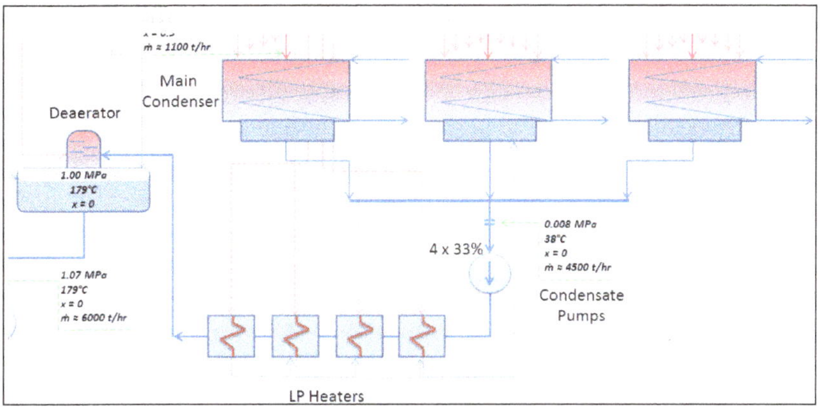

The main steam condenser (MC) system is designed to condense and deaerate the exhaust steam from the main turbine and provide a heat sink for the turbine bypass system. The exhausted steam from the LP turbines is condensed by passing over tubes containing water from the cooling system. There is a main condenser unit under each LP turbine, usually below the turbine with its axis perpendicular to the turbine axis. Since nuclear power plants usually contain also an auxiliary condenser (e.g. to condense steam from steam driven feedwater pumps), engineers use the term "main condenser".

The condenser must maintain a sufficient low vacuum in order to increase the power plant efficiency. The vacuum pumps maintain a sufficient vacuum in the condenser by extracting air and uncondensed gases. The lowest feasible condenser pressure is the saturation pressure corresponding to the ambient temperature (e.g. absolute pressure of 0.008 MPa, which corresponds to 41.5°C). Note that, there is always a temperature difference between (around $\Delta T = 14$°C) the condenser temperature and the ambient temperature, which originates from finite size and efficiency of condensers. Since neither the condenser is 100% efficient heat exchanger, there is always a temperature difference between the saturation temperature (secondary side) and the temperature of the coolant in the cooling system. Moreover, there is a design inefficiciency, which decreases the overall efficiency of the turbine. Ideally the steam exhausted into the condenser would have no subcooling. But real condensers are designed to subcool the liquid by a few degrees of Celsius in order to avoid the suction cavitation in the condensate pumps.

The steam condensers are broadly classified into two types:

- Surface condensers (or non-mixing type condensers). In surface condensers, there is no direct contact between the exhaust steam and the cooling water.
- Jet condensers (or mixing type condensers). In jet condensers there is direct contact between the exhaust steam and cooling water.

LEIDENFROST EFFECT

The Leidenfrost effect is a physical phenomenon in which a liquid, close to a surface that is significantly hotter than the liquid's boiling point, produces an insulating vapor layer that keeps the liquid from boiling rapidly. Because of this 'repulsive force', a droplet hovers over the surface rather than making physical contact with the hot surface.

This is most commonly seen when cooking, when a few drops of water are sprinkled in a hot pan. If the pan's temperature is at or above the Leidenfrost point, which is approximately 193 °C (379 °F) for water, the water skitters across the pan and takes longer to evaporate than it would take if the water droplets had been sprinkled into a cooler pan. The effect is responsible for the ability of a person to quickly dip a wet finger in molten lead, or blow out a mouthful of liquid nitrogen, without injury. The latter is potentially lethal, particularly should one accidentally swallow the liquid nitrogen.

The effect is named after Johann Gottlob Leidenfrost, who described it in A Tract About Some Qualities of Common Water in 1751.

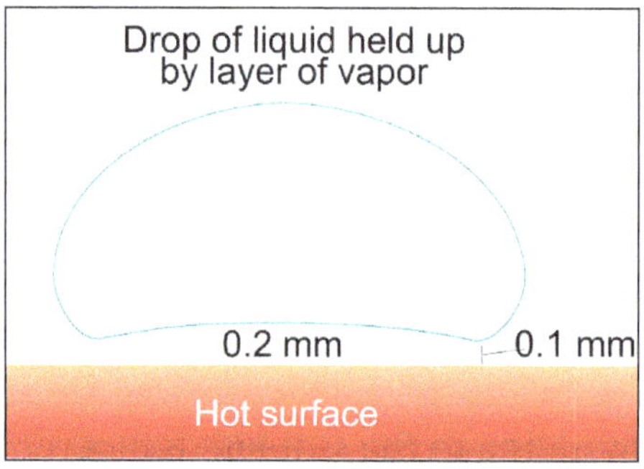

Leidenfrost droplet.

The effect can be seen as drops of water are sprinkled onto a pan at various times as it heats up. Initially, as the temperature of the pan is just below 100 °C (212 °F), the water flattens out and slowly evaporates, or if the temperature of the pan is well below 100 °C (212 °F), the water stays liquid. As the temperature of the pan goes above 100 °C (212 °F), the water droplets hiss when touching the pan and these droplets evaporate quickly. Later, as the temperature exceeds the Leidenfrost point, the Leidenfrost effect comes into play. On contact with the pan, the water droplets bunch up into small balls of water and skitter around, lasting much longer than when the temperature of the pan was lower. This effect works until a much higher temperature causes any further drops of water to evaporate too quickly to cause this effect.

This is because at temperatures above the Leidenfrost point, the bottom part of the water droplet vaporizes immediately on contact with the hot pan. The resulting gas suspends the rest of the water droplet just above it, preventing any further direct contact between the liquid water and the hot pan. As steam has much poorer thermal conductivity than the metal pan, further heat transfer between the pan and the droplet is slowed down dramatically. This also results in the drop being able to skid around the pan on the layer of gas just under it.

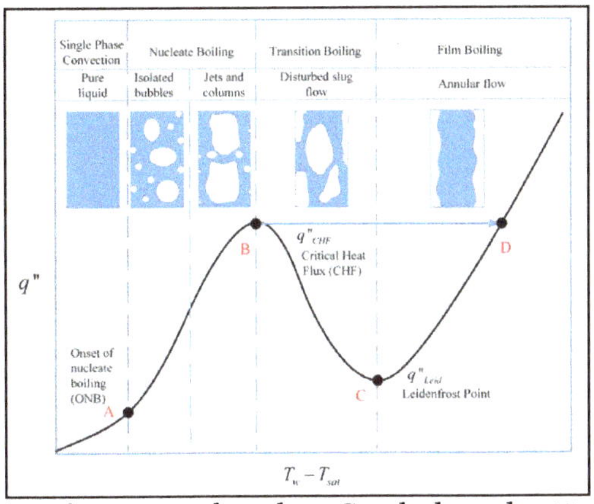

Behavior of water on a hot plate. Graph shows heat transfer (flux) vs temperature. Leidenfrost effect occurs after transition boiling.

The temperature at which the Leidenfrost effect begins to occur is not easy to predict. Even if the volume of the drop of liquid stays the same, the Leidenfrost point may be quite different, with a complicated dependence on the properties of the surface, as well as any impurities in the liquid. Some research has been conducted into a theoretical model of the system, but it is quite complicated. As a very rough estimate, the Leidenfrost point for a drop of water on a frying pan might occur at 193 °C (379 °F).

The effect was also described by the eminent Victorian steam boiler designer, Sir William Fairbairn, in reference to its effect on massively reducing heat transfer from a hot iron surface to water, such as within a boiler. On boiler design, he cited the work of Pierre Hippolyte Boutigny and Professor Bowman of King's College, London in studying this. A drop of water that was vaporized almost immediately at 168 °C (334 °F) persisted for 152 seconds at 202 °C (396 °F). Lower temperatures in a boiler firebox might evaporate water more quickly as a result; compare Mpemba effect. An alternative approach was to increase the temperature beyond the Leidenfrost point. Fairbairn considered this too, and may have been contemplating the flash steam boiler, but considered the technical aspects insurmountable for the time.

The Leidenfrost point may also be taken to be the temperature for which the hovering droplet lasts longest.

It has been demonstrated that it is possible to stabilize the Leidenfrost vapour layer of water by exploiting superhydrophobic surfaces. In this case, once the vapour layer is established, cooling never collapses the layer, and no nucleate boiling occurs; the layer instead slowly relaxes until the surface is cooled.

Leidenfrost effect has been used for the development of high sensitivity ambient mass spectrometry. Under the influence of Leidenfrost condition the levitating droplet does not release molecules out and the molecules are enriched inside the droplet. At the last moment of droplet evaporation all of the enriched molecules release in a short time domain and thus increase the sensitivity.

A heat engine based on the Leidenfrost effect has been prototyped. It has the advantage of extremely low friction.

Leidenfrost Point

A water droplet experiencing Leidenfrost effect on a hot stove plate.

The Leidenfrost point signifies the onset of stable film boiling. It represents the point on the boiling curve where the heat flux is at the minimum and the surface is completely covered by a vapor blanket. Heat transfer from the surface to the liquid occurs by conduction and radiation through the vapor. In 1756, Leidenfrost observed that water droplets supported by the vapor film slowly evaporate as they move about on the hot surface. As the surface temperature is increased, radiation through the vapor film becomes more significant and the heat flux increases with increasing excess temperature.

The minimum heat flux for a large horizontal plate can be derived from Zuber's equation,

$$\frac{q}{A}_{min} = Ch_{fg}\rho_v\left[\frac{\sigma g(\rho_L-\rho_v)}{(\rho_L+\rho_v)^2}\right]^{1/4}$$

where the properties are evaluated at saturation temperature. Zuber's constant, C is approximately 0.09 for most fluids at moderate pressures.

Heat Transfer Correlations

The heat transfer coefficient may be approximated using Bromley's equation,

$$h = C\left[\frac{k_v^3\rho_v g(\rho_L-\rho_v)\left(h_{fg}+0.4c_{pv}(T_s-T_{sat})\right)}{D_o\mu_v(T_s-T_{sat})}\right]^{1/4}$$

Where, D_o is the outside diameter of the tube. The correlation constant C is 0.62 for horizontal cylinders and vertical plates and 0.67 for spheres. Vapor properties are evaluated at film temperature.

For stable film boiling on a horizontal surface, Berenson has modified Bromley's equation to yield,

$$h = 0.425\left[\frac{k_{vf}^3 \rho_{vf} g(\rho_L - \rho_v)(h_{fg} + 0.4c_{pv}(T_s - T_{sat}))}{\mu_{vf}(T_s - T_{sat})\sqrt{\sigma / g(\rho_L - \rho_v)}}\right]^{1/4}$$

For vertical tubes, Hsu and Westwater have correlated the following equation:

$$h\left[\frac{\mu_v^2}{g\rho_v(\rho_L - \rho_v)k_v^3}\right]^{1/3} = 0.0020\left[\frac{4m}{\pi D_v \mu_v}\right]^{0.6}$$

Where, m is the mass flow rate in lb_m / hr at the upper end of the tube,

At excess temperatures above that at the minimum heat flux, the contribution of radiation becomes appreciable and becomes dominant at high excess temperatures. The total heat transfer coefficient can be is thus a combination of the two. Bromley has suggested the following equations for film boiling boiling from the outer surface of horizontal tubes.

$$h^{4/3} = h_{conv}^{4/3} + h_{rad} h^{4/3}$$

$$h_{rad} < h_{conv}$$

$$h = h_{conv} + \frac{3}{4} h_{rad}$$

The effective radiation coefficient, h_{rad} can be expressed as,

$$h_{rad} = \frac{\varepsilon\sigma(T_s^4 - T_{sat}^4)}{(T_s - T_{sat})}$$

Where, ε is the emissivity of the solid and σ is the Stefan-Boltzmann constant.

Pressure Field in a Leidenfrost Droplet

The equation for the pressure field in the vapor region between the droplet and the solid surface can be solved for using the standard momentum and continuity equations. For the sake of simplicity in solving, a linear temperature profile and a parabolic velocity profile are assumed within the vapor phase. The heat transfer within the vapor phase is assumed to be through conduction. With these approximations, the Navier-Stokes equation can be solved to get the pressure field.

Leidenfrost Temperature and Surface Tension Effects

The Leidenfrost temperature is the property of a given set of solid-liquid pair. The temperature of the solid surface beyond which the liquid undergoes Leidenfrost phenomenon is termed as Leidenfrost temperature. The calculation of Leidenfrost temperature involves the calculation of minimum film boiling temperature of a fluid. Berenson obtained a relation for the minimum film boiling temperature from minimum heat flux arguments. While the equation for the minimum film boiling temperature, which can be found in the reference above, is quite complex, the features of it can be understood from a physical perspective. One critical parameter to consider is the surface tension. The proportional relationship between the minimum film boiling temperature and surface tension is to be expected since fluids with higher surface tension need higher quantities of heat flux for the onset of nucleate boiling. Since film boiling occurs after nucleate boiling, the minimum temperature for film boiling should have a proportional dependence on the surface tension.

Henry developed a model for Leidenfrost phenomenon which includes transient wetting and microlayer evaporation. Since the Leidenfrost phenomenon is a special case of film boiling, the Leidenfrost temperature is related to the minimum film boiling temperature via a relation which factors in the properties of the solid being used. While the Leidenfrost temperature is not directly related to the surface tension of the fluid, it is indirectly dependent on it through the film boiling temperature. For fluids with similar thermophysical properties, the one with higher surface tension usually has a higher Leidenfrost temperature.

For example, for a saturated water-copper interface, the Leidenfrost temperature is 257 °C (495 °F). The Leidenfrost temperatures for glycerol and common alcohols are significantly smaller because of their lower surface tension values (density and viscosity differences are also contributing factors.)

Reactive Leidenfrost Effect

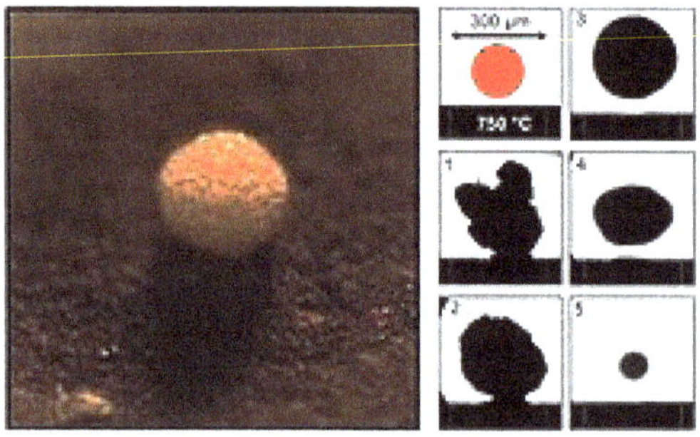

Reactive Leidenfrost effect of cellulose on silica, 750 °C (1,380 °F).

Non-volatile materials were discovered in 2015 to also exhibit a 'reactive Leidenfrost effect,' whereby solid particles were observed to float above hot surfaces and skitter around erratically. Detailed characterization of the reactive Leidenfrost effect was completed for small particles of cellulose (~0.5 mm) on high temperature polished surfaces by high speed photography. Cellulose was shown to decompose to short-chain oligomers which melt and wet smooth surfaces with increasing heat transfer associated with increasing surface temperature. Above 675 °C (1,247 °F), cellulose was observed to exhibit transition boiling with violent bubbling and associated reduction in heat transfer. Liftoff of the cellulose droplet (depicted at the right) was observed to occur above about 750 °C (1,380 °F) associated with a dramatic reduction in heat transfer.

High speed photography of the reactive Leidenfrost effect of cellulose on porous surfaces (macroporous alumina) was also shown to suppress the reactive Leidenfrost effect and enhance overall heat transfer rates to the particle from the surface. The new phenomenon of a 'reactive Leidenfrost (RL) effect' was characterized by a dimensionless quantity ($\varphi_{RL} = \tau_{conv}/\tau_{rxn}$), which relates the time constant of solid particle heat transfer to the time constant of particle reaction, with the reactive Leidenfrost effect occurring for $10^{-1} < \varphi_{RL} < 10^{+1}$. The reactive Leidenfrost effect with cellulose will occur in numerous high temperature applications with carbohydrate polymers including biomass conversion to biofuels, preparation and cooking of food, and tobacco use.

DROPWISE CONDENSATION

Dropwise condensation occurs when a vapor condenses on a surface not wetted by the condensate. For nonmetal vapors, dropwise condensation gives much higher heat transfer coefficients than those found with film condensation. For instance, the heat transfer coefficient for dropwise condensation of steam is around 10 times that for film condensation at power station condenser pressures and more than 20 times that for film condensation at atmospheric pressure. In circumstances where the filmwise coefficient is of similar magnitude to that on the cooling side, a change of mode to dropwise condensation offers a potential improvement in overall coefficient by a factor of up to around 2.

Clean metal surfaces are wetted by nonmetallic liquids and film condensation is the mode which normally occurs in practice. Nonwetting agents, known as dropwise promoters, are needed to promote dropwise condensation. Successful industrial application of dropwise condensation has been prevented by promoter breakdown, often associated with surface oxidation. Polytetrafluorethylene (ptfe, "teflon") provides an excellent nonwetting surface, but it has not been possible to produce sufficiently thin durable surface layers.

Various promoters have been identified and used successfully in laboratory tests, mainly with steam and also a few organic fluids having relatively high surface tension. A typical example is dioctadecyl disulphide, which has given lifetimes of hundreds or thousands of hours in laboratory investigations. Dropwise condensation of steam has been observed on chromium and gold surfaces which are very smooth, without use of a promoting agent. It seems probable, however, that nonwetting impurities were present.

The high heat transfer coefficients obtainable with dropwise condensation are very susceptible to reduction by the presence in the vapor of noncondensing gas. In the absence of significant vapor velocity, very small gas concentrations lead to appreciable lowering of the heat transfer coefficient. This was largely responsible for wide discrepancies between early published values.

During dropwise condensation, the bare surface is continually exposed to vapor by coalescences between drops and by the sweeping action of the falling drops as they are removed from the surface by gravity or vapor shear stress. "Primary" drops are formed at nucleation sites on the exposed surface (typical nucleation site densities are 107 to 108 sites/mm2). The primary drops grow by condensation until coalescences occur between neighbors. The coalesced drops continue to grow and new ones form and grow at sites exposed through coalescences. As the process continues, coalescences occur between drops of various sizes while the largest drops continue to grow until they reach their maximum size, when they are removed from the surface by gravity or vapor shear stress. The diameter ratio between the largest and smallest drops during dropwise condensation of steam is around 10^6.

Le Fevre and Rose have noted that three factors are involved in the mechanism of heat transfer through a single drop. These are: conduction in the liquid (important for relatively large drops), interphase matter transfer at the vapor-liquid interface (important for very small drops) and curvature of the vapor-liquid interface (important for the smallest drops).

By using an approximate expression for the distribution of drop sizes, together with their equation for the heat-transfer rate through a drop of given size, Le Fevre and Rose obtained the relation between the mean heat flux for the surface and the vapor-to-surface temperature difference. This showed, in agreement with experiment and in contrast to film condensation, that the heat transfer coefficient for dropwise condensation increases with increasing vapor-to-surface temperature difference.

Theory and experiment also indicate that the heat transfer coefficient decreases with decreasing pressure. Although in closed form and in principle applicable to any fluid, the expression giving the heat transfer coefficient is lengthy. An empirical equation in good agreement with both theory and experiment for dropwise condensation of pure, quiescent steam is

$$\frac{\alpha}{\mathrm{kW/m^2\,K}} = \theta^{0.8}\left(5 + 0.3\frac{\Delta T}{\mathrm{K}}\right)$$

Where α is the vapor-to-surface heat-transfer coefficient, ΔT is the vapor-to-surface temperature difference and θ is the Celsius temperature of the vapor.

There is conflicting evidence for the effect of the thermal conductivity of the condenser surface material on the heat-transfer coefficient resulting from non-uniformity of the surface heat flux. The balance of evidence suggests that this is only important at very low vapor-to-surface temperature difference where the condensing side resistance would probably be negligible in practice.

BOILING HEAT TRANSFER

Boiling heat transfer is heat transferred by the boiling of water. In BWRs boiling of coolant occurs at normal operation and it is very desired phenomenon. During an unusual, transient, large amounts of additional core boiling could occur and safety limits could be exceeded. Therefore, boiling heat transfer is an extremely important topic for accident considerations, for both BWRs and PWRs. Boiling heat transfer is not a fourth method of heat transfer, but it is considered separately because of its importance.

Categorization by the wall superheat temperature, ΔT_{sat}:

The pioneering work on boiling was done in 1934 by S. Nukiyama, who used electrically heated nichrome and platinum wires immersed in liquids in his experiments. Nukiyama was the first to identify different regimes of pool boiling using his apparatus. He noticed that boiling takes different forms, depending on the value of the wall superheat temperature ΔT_{sat} (known also as the excess temperature), which is defined as the difference between the wall temperature, T_{wall} and the saturation temperature, T_{sat}.

Four different boiling regimes of pool boiling (based on the excess temperature) are observed:

- Natural Convection Boiling $\Delta T_{sat} < 5°C$.
- Nucleate Boiling $5°C < \Delta T_{sat} < 30°C$.
- Transition Boiling $30°C < \Delta T_{sat} < 200°C$.
- Film Boiling $200°C < \Delta T_{sat}$.

Nucleate Boiling

The most common type of local boiling encountered in nuclear facilities is nucleate boiling. But in case of nuclear reactors, nucleate boiling occurs at significant flow rates through the reactor.

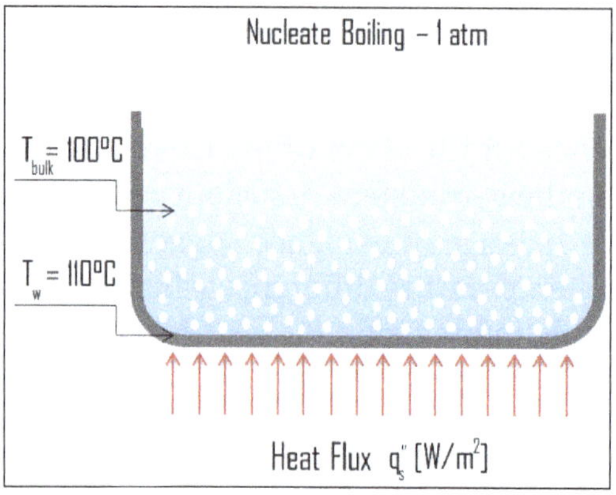

In nucleate boiling, steam bubbles form at the heat transfer surface and then break away and are carried into the main stream of the fluid. Such movement enhances heat transfer because the heat generated at the surface is carried directly into the fluid stream. Once in the main fluid stream, the bubbles collapse because the bulk temperature of the fluid is not as high as the heat transfer surface temperature where the bubbles were created. As was written, nucleate boiling at the surface effectively disrupts this stagnant layer and therefore nucleate boiling significantly improves the ability of a surface to transfer thermal energy to bulk fluid. This heat transfer process is sometimes desirable because the energy created at the heat transfer surface is quickly and efficiently "carried" away.

Close to the wall the situation is complex for several mechanisms increase the heat flux above that for pure conduction through the liquid.

- Note that, even in turbulent flow, there is a stagnant fluid film layer (laminar sublayer), that isolates the surface of the heat exchanger. The upward flux (due to buoyant forces) of vapor away from the wall must be balanced by an equal mass flux of liquid and this brings cooler liquid into closer proximity to the wall.
- The formation and movement of the bubbles turbulises the liquid near the wall and thus increases heat transfer from the wall to the liquid.
- Boiling differ from other forms of convection in that it depends on the latent heat of vaporization, which is very high for common pressures, therefore large amounts of heat can be transferred during boiling essentially at constant temperature.

The nucleate boiling heat flux cannot be increased indefinitely. At some value, we call it the "critical heat flux" (CHF), the steam produced can form an insulating layer over the surface, which in turn deteriorates the heat transfer coefficient. This is because a large fraction of the surface is covered by a vapor film, which acts as an thermal insulation due to the low thermal conductivity of the vapor relative to that of the liquid. Immediately after the critical heat flux has been reached, boiling become unstable and transition boiling occurs. The transition from nucleate boiling to film boiling is known as the "boiling crisis". Since beyond the CHF point the heat transfer coefficient decreases, the transition to film boiling is usually inevitable.

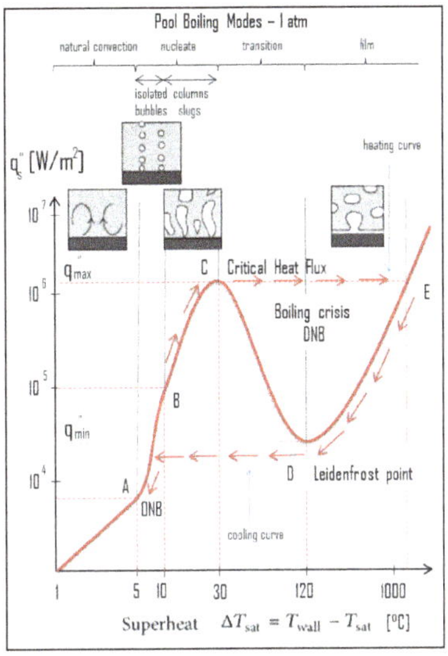

In following section, we will distinguish between:

- Nucleate pool boiling.
- Nucleate flow boiling.

Nucleate Boiling Correlations – Pool Boiling

Boiling regimes discussed above differ considerably in their character. There are also different correlations that describe the heat transfer.

Nucleate Pool Boiling

Rohsenow Correlation

The most widely used correlation for the rate of heat transfer in the nucleate pool boiling was proposed in 1952 by Rohsenow:

Rohsenow correlation,

$$\frac{Q}{A} = q = \left[\frac{C_1 \Delta T}{h_{fg} \Pr^n C_{sf}}\right]^3 \mu_1 h_{fg} \left[\frac{g(\rho_1 - \rho v)}{g_0 \sigma}\right]^{0.5}$$

Where,

- q – nucleate pool boiling heat flux [W/m²].
- c_1 — specific heat of liquid J/kg K.

- ΔT — excess temperature °C or K.
- h_{fg} – enthalpy of vaporization, J/kg.
- Pr — Prandtl number of liquid.
- n — experimental constant equal to 1 for water and 1.7 for other fluids.
- C_{sf} — surface fluid factor, for example, water and nickel have a C_{sf} of 0.006.
- μ_1 — dynamic viscosity of the liquid kg/m.s.
- g – gravitational acceleration m/s^2.
- g_0 — force conversion factor kgm/Ns^2.
- ρ_1 — density of the liquid kg/m^3.
- ρ_v — density of vapour kg/m^3.
- σ — surface tension-liquid-vapour interface N/m.

As can be seen, $\Delta T \propto (q)^{1/3}$. This very important proportionality shows increasing ability of interface to transfer heat.

Nucleate Boiling – Flow Boiling

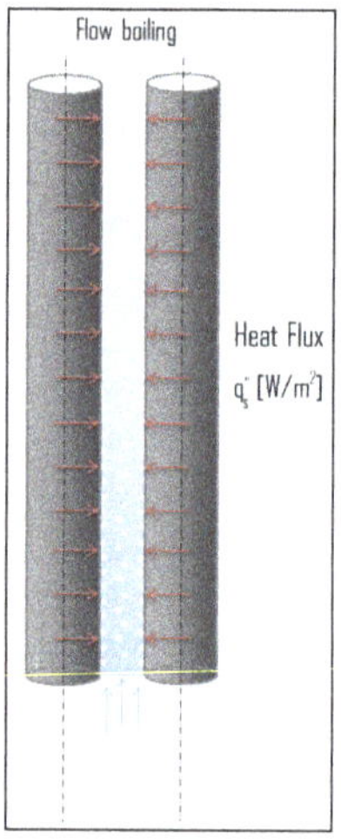

In flow boiling (or forced convection boiling), fluid flow is forced over a surface by external means such as a pump, as well as by buoyancy effects. Therefore, flow boiling is always accompanied by other convection effects. Conditions depend strongly on geometry, which may involve external flow over heated plates and cylinders or internal (duct) flow. In nuclear reactors, most of boiling regimes are just forced convection boiling. The flow boiling is also classified as either external and internal flow boiling depending on whether the fluid is forced to flow over a heated surface or inside a heated channel.

Internal flow boiling is much more complicated in nature than external flow boiling because there is no free surface for the vapor to escape, and thus both the liquid and the vapor are forced to flow together. The two-phase flow in a tube exhibits different flow boiling regimes, depending on the relative amounts of the liquid and the vapor phases. Therefore internal forced convection boiling is commonly referred to as two-phase flow.

Nucleate Boiling Correlations – Flow Boiling

McAdams Correlation

In fully developed nucleate boiling with saturated coolant, the wall temperature is determined by local heat flux and pressure and is only slightly dependent on the Reynolds number. For subcooled water at absolute pressures between 0.1 – 0.6 MPa, McAdams correlation gives:

$$q[Btu/hr.ft^2] = 0.074[T_w - T_{sat}]^{3.86}$$

$$^*[T_w - T_{sat}]\, is\, in\, Fahrenheits$$

Thom Correlation

The Thom correlation is for the flow boiling (subcooled or saturated at pressures up to about 20 MPa) under conditions where the nucleate boiling contribution predominates over forced convection. This correlation is useful for rough estimation of expected temperature difference given the heat flux:

$$T_w - T_{sat} = 22.5\, q^{0.5} . e^{-p/8.7}$$

where:

q is the heat flex MW/m^2

p is the pressure MP_a

Chen's Correlation

In 1963, Chen proposed the first flow boiling correlation for evaporation in vertical tubes to attain widespread use. Chen's correlation includes both the heat transfer coefficients due to nucleate boiling as well as forced convective mechanisms. It must be noted, at higher vapor fractions, the heat transfer coefficient varies strongly with flow rate. The flow velocity in a core can be very high causing very high turbulences. This heat transfer mechanism has been referred to as "forced convection evaporation". No adequate criteria has been established to determine the transition from nucleate boiling to forced convection vaporization. However, a single correlation that is valid for both nucleate boiling and forced convection vaporization has been developed by Chen for saturated boiling conditions and extended to include subcooled boiling by others. Chen proposed a correlation where the heat transfer coefficient is the sum of a forced convection component and a nucleate boiling component. It must be noted, the nucleate pool boiling correlation of Forster and Zuber is used to calculate the nucleate boiling heat transfer coefficient, h_{FZ} and the turbulent flow correlation of Dittus-Boelter is used to calculate the liquid-phase convective heat transfer coefficient, h_l.

Chen's Correlation

$$h_{tp} = h_{FZ} S + h_l F$$

$$h_{tp} = S \overbrace{\frac{K_l^{0.79} c_l^{0.45} \rho_l^{0.49} g_c^{0.25} \Delta T^{0.24} \Delta P^{0.75}}{\sigma^{0.5} \mu_l^{0.29} H_{fg}^{0.29} \rho_v^{0.24}}}^{nucleate\ pool\ boiling\ Forster - Zuber} + \overbrace{F \, \mathrm{Re}_l^{0.8} \, \mathrm{Pr}_l^{0.4} \, k_l / e}^{force\ convection\ Dittus - Boelter}$$

$$S = \frac{1}{1 + 0 \cdot 00000253 \, \mathrm{Re}_{tp}^{1 \cdot 17}}$$

$$F = \left(\frac{1}{X_n} + 0 \cdot 213 \right)^{0 \cdot 736}$$

Where,

- H_{tp} – two phase heat transfer coefficient.
- S – the nucleate boiling suppression factor; 0.122 for FZ correlation.
- F – two-phase multiplier; correction factor which is efine as the ratio of the true two-phase Reynols number to the single-phase.
- K_l- thermal con uctivity of liquid.
- C_l – specific heat of liquid.
- G_c – gravitational conversion factor.
- ΔT -wall superheat.
- ΔP – ifference in saturation an wall superheat pressures.
- σ – surface tension.
- μ – viscosity.
- H_{fg} – enthalpy of vaporization.
- E – equivent iameter.

The nucleate boiling suppression factor, S, is the ratio of the effective superheat to wall superheat. It accounts for decreased boiling heat transfer because the effective superheat across the boundary layer is less than the superheat based on wall temperature. The two-phase multiplier, F, is a function of the Martinelli parameter χ_{tt}.

Boiling Crisis – Critical Heat Flux

As was written, in nuclear reactors, limitations of the local heat flux is of the highest importance for reactor safety. For pressurized water reactorsand also for boiling water reactors, there are thermal-hydraulic phenomena, which cause a sudden decrease in the efficiency of heat transfer (more precisely in the heat transfer coefficient). These phenomena occur at certain value of heat flux, known as the "critical heat flux". The phenomena, that cause the deterioration of heat transfer are different for PWRs and for BWRs.

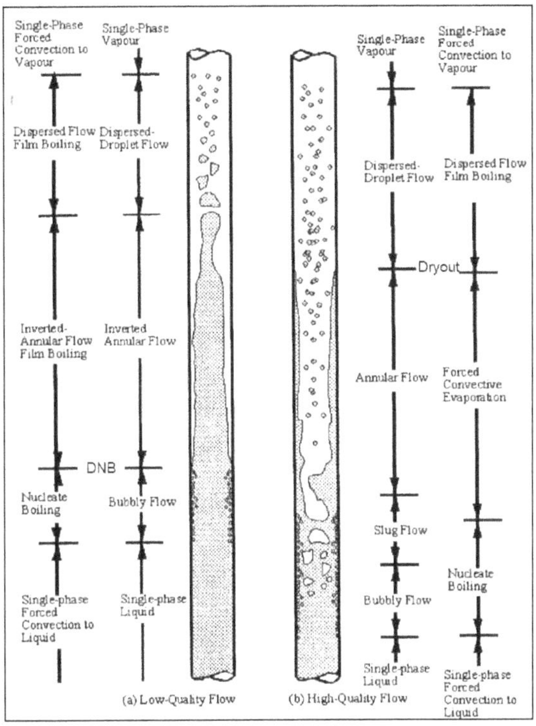

In both types of reactors, the problem is more or less associated with departure from nucleate boiling. The nucleate boiling heat flux cannot be increased indefinitely. At some value, we call it the "critical heat flux" (CHF), the steam produced can form an insulating layer over the surface, which in turn deteriorates the heat transfer coefficient. Immediately after the critical heat flux has been reached, boiling become unstable and film boiling occurs. The transition from nucleate boiling to film boiling is known as the "boiling crisis". As was written, the phenomena, that cause the deterioration of heat transfer are different for PWRs and for BWRs.

Departure From Nucleate Boiling – DNB

In case of PWRs, the critical safety issue is named DNB(departure from nucleate boiling), which causes the formation of a local vapor layer, causing a dramatic reduction in heat transfer capability. This phenomenon occurs in the subcooled or low-quality region. The behaviour of the boiling crisis depends on many flow conditions (pressure, temperature, flow rate), but the boiling crisis occurs at a relatively high heat fluxes and appears to be associated with the cloud of bubbles, adjacent to the surface. These bubbles or film of vapor reduces the amount of incoming water. Since this phenomenon deteriorates the heat transfer coefficient and the heat flux remains, heat then accumulates in the fuel rod causing dramatic rise of cladding and fuel temperature. Simply, a very high temperature difference is required to transfer the critical heat flux being produced from the surface of the fuel rod to the reactor coolant (through vapor layer).

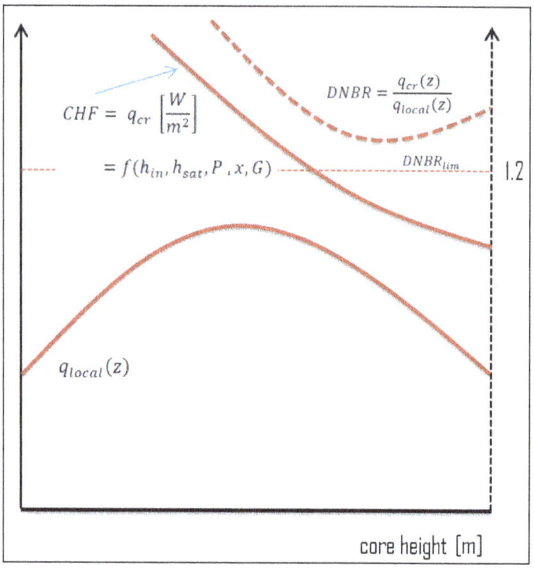

In case of PWRs, the critical flow is inverted annular flow, while in BWRs, the critical flow is usually annular flow. The difference in flow regime between post-dryout flow and post-DNB flow is depicted in the figure. In PWRs at normal operation the flow is considered to be single-phase. But a great deal of study has been performed on the nature of two-phase flow in case of transients and accidents (such as the loss-of-coolant accident – LOCA or trip of RCPs), which are of importance in reactor safety and in must be proved and declared in the Safety Analysis Report (SAR).

In pressurized water reactors, one of key safety requirements is that a departure from nucleate boiling (DNB) will not occur during steady state operation, normal operational transients, and anticipated operational occurrences (AOOs). Fuel cladding integrity will be maintained if the minimum DNBR remains above the 95/95 DNBR limit for PWRs (a 95% probability at a 95% confidence level). DNB criterion is one of acceptance criteria in safety analyses as well as it constitutes one of safety limits in technical specifications.

An important duty of the plant operator is to control plant parameters such that a safe margin to DNB (or distance from DNB on the heat transfer curve) is maintained. Any sudden, large change in the following plant parameters/directions will decrease the margin to DNB:

- Decrease in reactor coolant pressure.
- Decrease in reactor coolant flow rate.

- Increase in reactor power.
- Increase in reactor coolant inlet temperature.

Therefore, the function of the operators and the plant design is to prevent a sudden, large change in these plant parameters.

Dryout – BWRs

In BWRs, similar phenomenon is known as "dryout" and it is directly associated with changes in flow pattern during evaporation in the high-quality region. At given combinations of flow rate through a channel, pressure, flow quality, and linear heat rate, the wall liquid film may exhaust and the wall may be dried out. At normal, the fuel surface is effectively cooled by boiling coolant. However when the heat flux exceeds a critical value (CHF – critical heat flux) the flow pattern may reach the dryout conditions (thin film of liquid disappears). The heat transfer from the fuel surface into the coolant is deteriorated, with the result of a drastically increased fuel surface temperature. In the high-quality region, the crisis occurs at a lower heat flux. Since the flow velocity in the vapor core is high, post-CHF heat transfer is much better than for low-quality critical flux (i.e. for PWRs temperature rises are higher and more rapid).

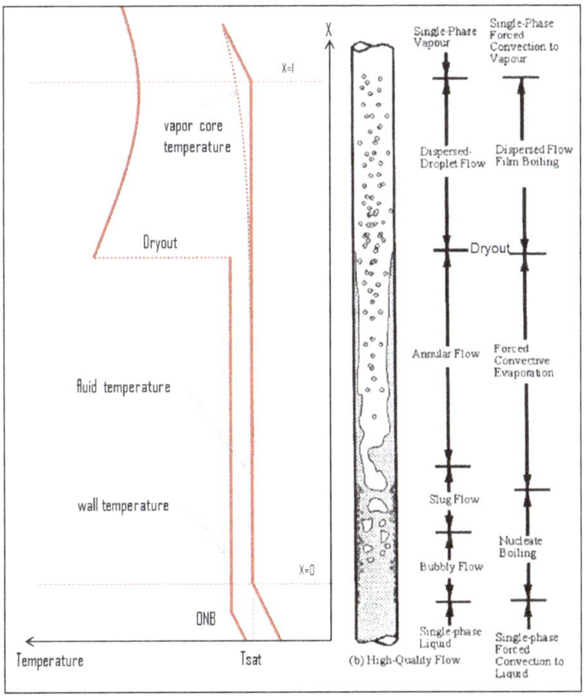

Nuclear reactors produce enormous amount of heat (energy) in a small volume. The density of the energy generation is very large and this puts demands on its heat transfer system (reactor coolant system). For a reactor to operate in a steady state, all of the heat released in the system must be removed as fast as it is produced. This is accomplished by passing a liquid or gaseous coolant through the core and through other regions where heat is generated. The heat transfer must be equal to or greater than the heat generation rate or overheating and possible damage to the fuel may occur.

The temperature in an operating reactor varies from point to point within the system. As a consequence, there is always one fuel rod and one local volume, that are hotter than all the rest. In order to limit these hot places the peak power limits must be introduced. The peak power limits are associated with such phenomena as the departure from nucleate boiling and with the conditions which could cause fuel pellet melt.

Therefore power distribution within the core must be properly limited. These limitations are usually divided into two basic categories:

- Limitation of global power distribution:
    - Axial flux difference.
    - Power tilt.
- Limitation of local power distribution:
    - Local Heat Flux.
    - Enthalpy Rise in Hot Channel.

Nucleate Boiling

Nucleate boiling is a type of boiling that takes place when the surface temperature is hotter than the saturated fluid temperature by a certain amount but where the heat flux is below the critical heat flux. For water, as shown in the graph, nucleate boiling occurs when the surface temperature is higher than the saturation temperature (T_S) by between 10 °C (18 °F) to 30 °C (54 °F). The critical heat flux is the peak on the curve between nucleate boiling and transition boiling. The heat transfer from surface to liquid is greater than that in film boiling.

Mechanism

Two different regimes may be distinguished in the nucleate boiling range. When the temperature difference is between approximately 4 °C (7.2 °F) to 10 °C (18 °F) above T_S, isolated bubbles form at nucleation sites and separate from the surface. This separation induces considerable fluid mixing near the surface, substantially increasing the convective heat transfer coefficient and the heat flux. In this regime, most of the heat transfer is through direct transfer from the surface to the liquid in motion at the surface and not through the vapor bubbles rising from the surface.

Between 10 °C (18 °F) and 30 °C (54 °F) above T_S, a second flow regime may be observed. As more nucleation sites become active, increased bubble formation causes bubble interference and coalescence. In this region the vapor escapes as jets or columns which subsequently merge into slugs of vapor.

Interference between the densely populated bubbles inhibits the motion of liquid near the surface. This is observed on the graph as a change in the direction of the gradient of the curve or an inflection in the boiling curve. After this point, the heat transfer coefficient starts to reduce as the surface temperature is further increased although the product of the heat transfer coefficient and the temperature difference (the heat flux) is still increasing.

When the relative increase in the temperature difference is balanced by the relative reduction in the heat transfer coefficient, a maximum heat flux is achieved as observed by the peak in the graph. This is the critical heat flux. At this point in the maximum, considerable vapor is being formed, making it difficult for the liquid to continuously wet the surface to receive heat from the surface. This causes the heat flux to reduce after this point. At extremes, film boiling commonly known as the Leidenfrost effect is observed.

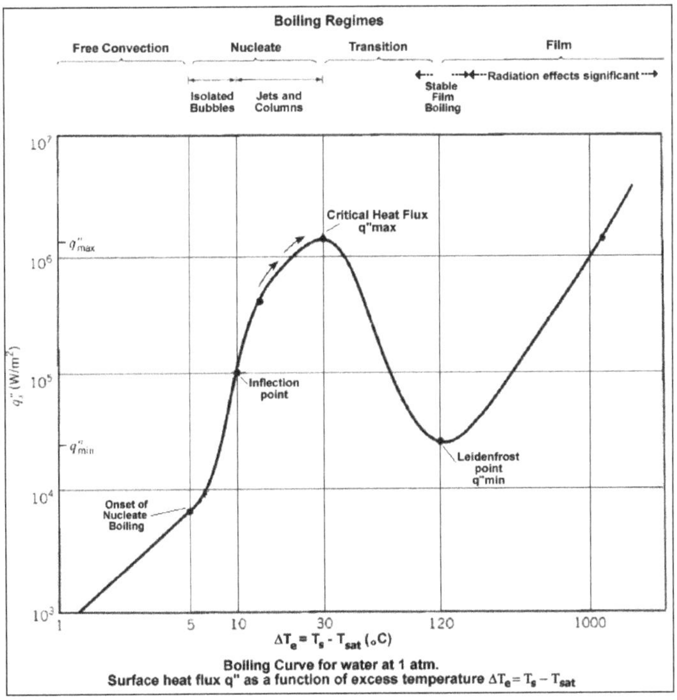

Boiling curve for water at 1atm.

The process of forming steam bubbles within liquid in micro cavities adjacent to the wall if the wall temperature at the heat transfer surface rises above the saturation temperature while the bulk of the liquid (heat exchanger) is subcooled. The bubbles grow until they reach some critical size, at which point they separate from the wall and are carried into the main fluid stream. There the bubbles collapse because the temperature of bulk fluid is not as high as at the heat transfer surface, where the bubbles were created. This collapsing is also responsible for the sound a water kettle produces during heat up but before the temperature at which bulk boiling is reached.

Heat transfer and mass transfer during nucleate boiling has a significant effect on the heat transfer rate. This heat transfer process helps quickly and efficiently to carry away the energy created at the heat transfer surface and is therefore sometimes desirable—for example in nuclear power plants, where liquid is used as a coolant.

The effects of nucleate boiling take place at two locations:

- The liquid-wall interface.
- The bubble-liquid interface.

The nucleate boiling process has a complex nature. A limited number of experimental studies provided valuable insights into the boiling phenomena, however these studies provided often contradictory data due to internal recalculation (state of chaos in the fluid not applying to classical thermodynamic methods of calculation, therefore giving wrong return values) and have not provided conclusive findings yet to develop models and correlations. Nucleate boiling phenomenon still requires more understanding.

Boiling Heat Transfer Correlations

The nucleate boiling regime is important to engineers because of the high heat fluxes possible with moderate temperature differences. The data can be correlated by equation of the form,

$$Nu_b = C_{fc}\left(Re_b, Pr_L\right)$$

The Nusselt number is defined as,

$$Nu_b = \frac{\left(\frac{q}{A}\right)D_b}{\left(T_s - T_{sat}\right)k_L}$$

where, q/A is the total heat flux, D_b is the maximum bubble diameter as it leaves the surface, $T_s - T_{sat}$ is the excess temperature, k_L is the thermal conductivity of the liquid and Pr_L is the Prandtl number of the liquid. The bubble Reynolds number, Re_b is defined as,

$$Re_b = \frac{D_b G_b}{\mu_L}$$

Where, G_b is the average mass velocity of the vapor leaving the surface and μL is the liquid viscosity.

Rohsenow has developed the first and most widely used correlation for nucleate boiling,

$$\frac{q}{A} = \mu_L h_{fg} \left[\frac{g(\rho_L - \rho_v)}{\sigma} \right]^{1/4} \left[\frac{c_{pL}(T_s - T_{sat})}{C_{sf} h_{fg} Pr_L^n} \right]^3$$

Where, C_pL is the specific heat of the liquid. C_{sf} is the surface fluid combination and vary for various combinations of fluid and surface. The variable n depends on the surface fluid combination and typically has a value of 1.0 or 1.7. For example, water and nickel have a C_{sf} of 0.006 and n of 1.0.

Values of C_{sf} for various surface fluid combinations	
Surface fluid combinations	C_{sf}
Water/copper	0.013
Water/nickel	0.006
Water/platinum	0.013
Water/brass	0.006
Water/stainless steel, mechanically polished	0.0132
Water/stainless steel, chemically etched	0.0133
Water/stainless steel, ground and polished	0.0080
CCl_4/copper	0.013
Benzene/chromium	0.0101
n-Pentane/chromium	0.015
Ethyl alcohol/chromium	0.0027
Isopropyl alcohol/copper	0.0025
n-Butyl alcohol/copper	0.003

Departure from Nucleate Boiling

If the heat flux of a boiling system is higher than the critical heat flux (CHF) of the system, the bulk fluid may boil, or in some cases, *regions* of the bulk fluid may boil where the fluid travels in small channels. Thus large bubbles form, sometimes blocking the passage of the fluid. This results in a departure from nucleate boiling (DNB) in which steam bubbles no longer break away from the solid surface of the channel, bubbles dominate the channel or surface, and the heat flux dramatically decreases. Vapor essentially insulates the bulk liquid from the hot surface.

During DNB, the surface temperature must therefore increase substantially above the bulk fluid temperature in order to maintain a high heat flux. Avoiding the CHF is an engineering problem in heat transfer applications, such as nuclear reactors, where fuel plates must not be allowed to overheat. DNB may be avoided in practice by increasing the pressure of the fluid, increasing its flow rate, or by utilizing a lower temperature bulk fluid which has a higher CHF. If the bulk fluid temperature is too low or the pressure of the fluid is too high, nucleate boiling is however not possible.

DNB is also known as transition boiling, unstable film boiling, and partial film boiling. For water boiling as shown on the graph, transition boiling occurs when the temperature difference between the surface and the boiling water is approximately 30 °C (54 °F) to 120 °C (220 °F) above the T_S. This corresponds to the high peak and the low peak on the boiling curve. The low point between transition boiling and film boiling is the Leidenfrost point.

During transition boiling of water, the bubble formation is so rapid that a vapor film or blanket begins to form at the surface. However, at any point on the surface, the conditions may oscillate between film and nucleate boiling, but the fraction of the total surface covered by the film increases with increasing temperature difference. As the thermal conductivity of the vapor is much less than that of the liquid, the convective heat transfer coefficient and the heat flux reduces with increasing temperature difference.

References

- Film-condensation, boiling-and-condensation, heat-transfer, nuclear-engineering: nuclear-power.net, Retrieved 10 July, 2019
- Bernardin, john d.; mudawar, issam (2002). "a cavity activation and bubble growth model of the leidenfrost point". Journal of heat transfer. 124 (5): 864–74. Doi:10.1115/1.1470487
- Boiling-heat-transfer, boiling-and-condensation, heat-transfer, nuclear-engineering: nuclear-power.net, Retrieved 13 March, 2019
- Vakarelski, ivan u.; patankar, neelesh a.; marston, jeremy o.; chan, derek y. C.; thoroddsen, sigurdur t. (2012). "stabilization of leidenfrost vapour layer by textured superhydrophobic surfaces". Nature. 489 (7415): 274–7. Bibcode:2012natur.489..274v. Doi:10.1038/nature11418. Pmid 22972299
- Boiling-crisis-critical-heat-flux, boiling-and-condensation, nuclear-engineering: nuclear-power.net, Retrieved 29 June, 2019
- Wells, gary g.; ledesma-aguilar, rodrigio; mchale, glen; sefiane, khellil (3 march 2015). "a sublimation heat engine". Nature communications. 6: 6390. Bibcode:2015natco...6e6390w. Doi:10.1038/ncomms7390. Pmc 4366496. Pmid 25731669

5

Mass Transfer

The net movement of mass from one location to another occurring in different processes like absorption and distillation is called mass transfer. The key concepts related to mass transfer are mass transfer coefficient, mass flux and chemical equilibrium. This chapter has been carefully written to provide an easy understanding of these facets of mass transfer.

The theory of mass transfer allows for the computation of mass flux in a system and the distribution of the mass of different species over time and space in such a system, also when chemical reactions are present. The purpose of such computations is to understand, and possibly design or control, such a system.

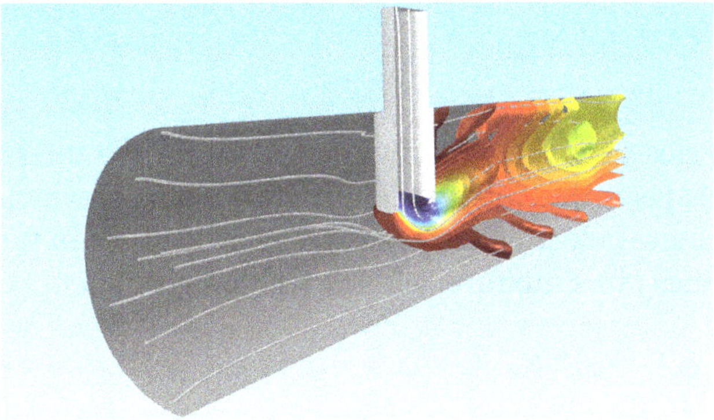

Transport and reactions in a reactor. The concentration isosurfaces reveal mass transfer through diffusion and convection. The flux through diffusion takes place perpendicular to the concentration isosurfaces, i.e., the reactions may cause a flux to the reaction site of the species that are consumed in the reaction. Convection creates a larger separation between the concentration isosurfaces and takes place along the streamlines of the fluid flow (in white), which in some places run along the isosurfaces, since convection tends to eliminate concentration gradients along its main direction.

Mathematical Description of Mass Transfer

The driving force, F, for mass transfer is created by gradients in the system potential, U:

$$F=-\nabla U$$

Gradients in chemical composition are usually responsible for this driving force. The driving force for transport over phase boundaries is generated by a deviation from equilibrium over such a phase boundary. Additional driving forces may contribute with a drift velocity, such as the forces created by migration, pressure, gravitational, and centrifugal forces.

The equation below shows the forces acting on a chemical species, per mole of atoms, ions, or molecules, due to gradients in chemical potential and electric fields (migration).

$$\mathrm{F}_i = \underbrace{-RT\nabla \ln a_i}_{\text{CHEMICAL}} \underbrace{- z_i F \nabla \phi}_{\text{ELECTRICAL}}$$

In these equations, R denotes the gas constant, T is the temperature, a_i is the activity of each species, z_i denotes the charge number of a species, F is the Faraday constant, and φ is the electric potential. The negative gradient of φ is the electric field. The activity can be understood as a thermodynamic measure of the chemical potential of the system, so that gradients in activity correspond to driving forces for chemical mass transport.

A simple chemical assumption is that the activity of a species i is given by its mole fraction, denoted as x_i. This is exactly true for an ideal mixture.

$$\mathrm{F}_\mathrm{i} = -RT\frac{1}{x_i}\nabla x_i - z_i F \nabla \phi$$

The forces on a species i are balanced by the friction in the interaction between this species and the other species in a mixture. The friction acting on a mole of i is proportional to the difference in mass velocity between i and each species j in the mixture, the mole fraction of each species j in a mixture, and the friction coefficient between i and j.

$$\mathrm{F}_{\mathrm{fric,i}} = -\sum_{\mathrm{j \neq i}} \varsigma_{ij} x_j (u_{R,i} - u_{R,j})$$

In this equation, ζ_{ij} denotes the friction coefficient between species i and j, x_j is the mole fraction of species j, and $\mathrm{u}_{\mathrm{R},i}$ is the mass species velocity of species i relative to the mass average velocity of the whole mixture. Note that the mass velocities of each species in the equation above are given using the mass average velocity for the mixture as reference. A species that does not deviate from the velocity of the mixture (i.e., that does not diffuse or migrate, in this case), has a zero $\mathrm{u}_{\mathrm{R},i}$, when using the mixture velocity as reference.

If we now set the driving forces to exactly balance the friction forces acting on species i, we obtain the following equation:

$$-RT\frac{1}{x_i}\nabla x_i - z_i F \nabla \phi - \sum_{\mathrm{j \neq i}} \varsigma_{ij} x_j (u_{R,i} - u_{R,j}) = 0$$

The molar flux is defined as,

$$J_i = c x_i u_{R,i}$$

where J_i is the flux vector of species *i* relative to the velocity of the mixture and *c* is the total concentration of all species in a mixture. Introducing the Maxwell-Stefan diffusivity as:

$$\mathcal{D}_{ij} = \frac{1}{\varsigma_{ij} RT}$$

and using the molar flux to eliminate $u_{R,i}$ and $u_{R,j}$ in the force balance equation above yields the following expression:

$$-c\nabla x_i - \frac{z_i F_C}{RT} x_i \nabla\phi = \sum_{j\neq i} \frac{1}{\mathcal{D}_{ij}} (x_j J_i - x_j J_i)$$

This is the *Maxwell-Stefan equation,* an equation that forms the basis for the mathematical description of mass transfer of chemical species in a mixture [2]. Simplifications of these equations for diluted mixtures give, for example, Fick's first law of diffusionand the Nernst-Planck equations for diffusion and migration.

The molar flux of a species *i* relative to a fixed coordinate system, denoted as N_i, is obtained by adding the convective term, due to the velocity of the whole mixture:

$$N_i = J_i + c_i u$$

The resulting fluxes are used in the mass conservation equations for each species in the solution:

$$\frac{\partial M_i c_i}{\partial t} + \nabla . M_i N_i - M_i R_i = 0$$

The sum of all mass fluxes, including the convective term, results in the continuity equation for the mixture:

$$\frac{\partial}{\partial t} = \underbrace{\sum_i M_i c_i}_{=\rho} + \nabla . \underbrace{\sum_i M_i N_i}_{=\rho u} - \underbrace{\sum_i M_i R_i}_{=0} = 0$$

where the last term is necessarily zero due to the conservation of mass of an individual chemical reaction. By identifying the sums as the density and mass flux density, we get the mass continuity equation:

$$\frac{\partial \rho}{\partial t} + \nabla . \rho u = 0$$

The convective term in the flux is the contribution to the flux of a species due to the movement of the whole solution. For this reason, convective flux takes place along the velocity streamlines of the solution for all chemical species in the solution. Note that the sum of all the species' mass fluxes, relative to the flux of the mixture, is zero if the mass-averaged velocity is used as reference. The mass-averaged velocity is defined as:

$$u = \frac{\sum_{i} \rho_i u_i}{\sum_{i} \rho_i}$$

where ∂_i denotes the mass density of a species *i*. This implies that, in general, the mass flux of each species is tightly coupled to the total mass velocity in a mixture. In a strict definition, the mass average velocity of a mixture could be obtained by formulating and solving the equations for the conservation of momentum for each species in a mixture.

However, the interaction coefficients, required for such formulations, are usually difficult to measure or calculate. Instead, equations for the conservation of momentum for the whole mixture are usually defined. The combination of the equations for the conservation of momentum and mass for a mixture at low velocities (less than one third of the speed of sound), yield the Navier-Stokes equations. The solution of the Navier-Stokes equations gives the velocity field (a vector field) that also determines the direction of the convective flux of all species in the mixture.

The tight coupling between mass transport for each species and the conservation of mass for the whole mixture is exemplified in the example below. Oxygen in air is consumed at the surface of a catalyst and produces liquid water that is removed from the gas phase in a gas diffusion electrode. The consumption of oxygen causes a net velocity in the gas mixture (air). Additionally, a nitrogen concentration gradient is formed in order to perfectly balance the advective (or convective) flux of nitrogen with an opposite flux by diffusion.

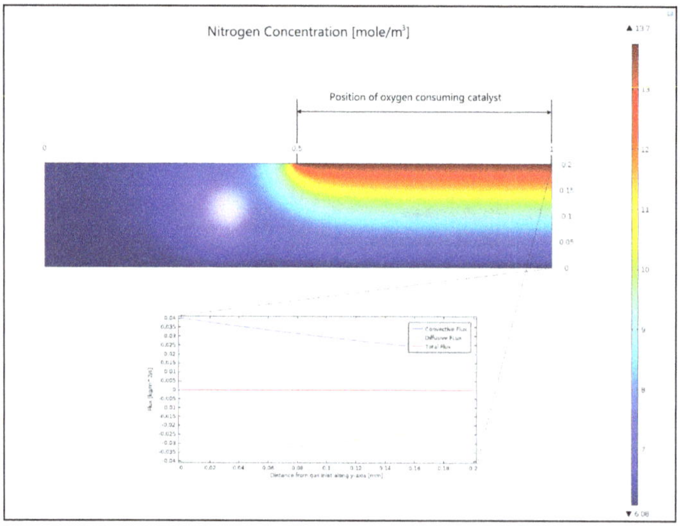

Surface plot of the nitrogen concentration in a gas diffusion electrode. The flux along the right vertical edge, plotted in the x-y plot, shows that the diffusive flux from the catalytic surface exactly compensates the convective flux to this surface, generated by oxygen consumption at the catalyst's surface.

Due to the difficulties of numerically resolving steep gradients in potential, mass transfer across phase boundaries is often expressed using difference equations, instead of differential equations. This approximation implies that the gradients, included in the driving forces, are linearized inside a fictitious boundary layer. The thickness of the layer is then defined as the distance from a phase boundary where the linearized concentration gradient, starting from the concentration at the phase boundary, reaches the bulk concentration. The definition of the boundary layer also means that its thickness may be different for different species.

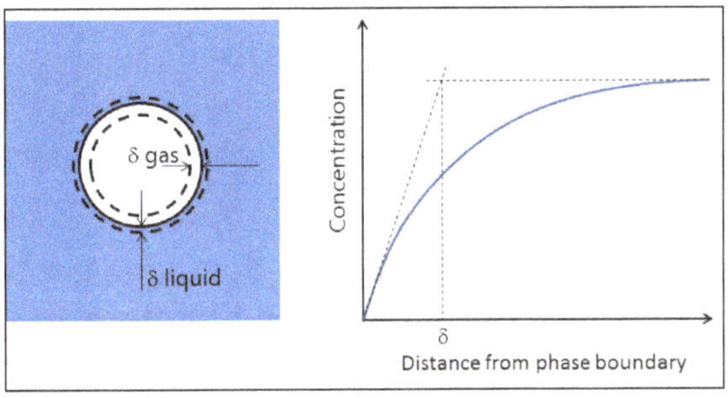

Fictive boundary layers inside and around a gas bubble in a liquid.

The mass transfer coefficient, k_m, for such an interface is defined as the diffusivity divided by the boundary layer thickness, δ.

$$k_m = \frac{D}{\delta}$$

An estimate of the relation between the boundary layer thickness and a system's typical length is given by the Sherwood number:

$$\text{Sh} = \frac{k_m L}{D}$$

In this expression, L denotes a typical length for a system, such as the radius of a pipe and the width of a channel. However, if the mass transport around a gas bubble in a liquid is studied, then Lmay denote the radius of the bubble. Since the thickness of the boundary layer depends on the convection just outside an interface, the Sherwood number also gives a measurement of the convective and diffusive fluxes to such an interface.

The boundary layer thickness at a liquid-gas interface for a rising bubble is of the order of magnitude of 100 μm in the gas phase and around 10 μm in the liquid phase.

The Sherwood number can also be defined as a function of the Reynolds and Schmidt numbers. The Reynolds number gives an estimate of the ratio of momentum transport by inertia to viscosity in a fluid:

$$\mathrm{Re} = \frac{\rho U L}{\mu}$$

where μ denotes viscosity and U denotes an average velocity. The Schmidt number gives an estimate of the relation between viscosity and diffusivity in a fluid:

$$Sc = \frac{\mu}{\rho D}$$

The mass transfer coefficient can be estimated from the relation between the Sherwood number and the Reynolds and Schmidt numbers. For example, for forced convection along a flat plate, the following expression can be used:

$$\mathrm{Sh} = 0.5\, f_{loc}\, \mathrm{Re}\, Sc^{\frac{1}{3}}$$

Where f_{loc} denotes the local friction factor for flow along a flat plate. The friction coefficient for different geometries is tabulated in literature and may also be obtained experimentally. All the material properties and the average velocity in the relation above are relatively easy to find in literature or estimate from simple calculations. Once the Sherwood number is calculated, then the mass transfer coefficient can be calculated, including the boundary layer thickness, which is the parameter that is not easily estimated otherwise. Note, though, that the above expression is only valid for flat plates.

MASS TRANSFER COEFFICIENT

In engineering, the mass transfer coefficient is a diffusion rate constant that relates the mass transfer rate, mass transfer area, and concentration change as driving force:

$$k_c = \frac{\dot{n}_A}{A \Delta c_A}$$

Where,

- k_c is the mass transfer coefficient [mol/(s·m²)/(mol/m³)], or m/s.
- $\dot{n}_A$ is the mass transfer rate [mol/s].
- A is the effective mass transfer area [m²].
- Δc_A is the driving force concentration difference [mol/m³].

This can be used to quantify the mass transfer between phases, immiscible and partially miscible fluid mixtures (or between a fluid and a porous solid). Quantifying mass transfer allows for design and manufacture of separation process equipment that can meet specified requirements, estimate what will happen in real life situations (chemical spill), etc.

Mass transfer coefficients can be estimated from many different theoretical equations, correlations, and analogies that are functions of material properties, intensive properties and flow regime (laminar or turbulent flow). Selection of the most applicable model is dependent on the materials and the system, or environment, being studied.

Mass Transfer Coefficient Units

- $(mol/s)/(m^2 \cdot mol/m^3) = m/s$

Note, the units will vary based upon which units the driving force is expressed in. The driving force shown here as 'Δc_A'' is expressed in units of moles per unit of volume, but in some cases the driving force is represented by other measures of concentration with different units. For example, the driving force may be partial pressures when dealing with mass transfer in a gas phase and thus use units of pressure.

CRITERION FOR CHEMICAL EQUILIBRIUM

The general criterion for thermodynamic equilibrium was derived, is section as:

$$(dG^2)_{T.P} \leq 0$$

the above equation implies that if a closed system undergoes a process of change while being under thermal and mechanical equilibrium, for all incremental changes associated with the compositions of each species, the total Gibbs free energy of the system would decrease. At complete equilibrium the equality sign holds; or, in other words, the Gibbs free energy of the system corresponds to the minimum value possible under the constraints of constant (and uniform) temperature and pressure. Since the criterion makes no assumptions as to the nature of the system in terms of the number of species or phases, or if reactions take place between the species, it may also be applied to determine a specific criterion for a reactive system under equilibrium.

At the initial state of a reaction, when the reactants are brought together a state of non-equilibrium ensues as reactants begin undergoing progressive transformation to products. However, a state of equilibrium must finally attain when the rates of forward and backward reactions equalize. Under such a condition, no further change in the composition of the residual reactants or products formed occurs. However, if we consider this particular state, we may conclude that while in a macroscopic sense the system is in a state of static equilibrium, in the microscopic sense there is dynamic equilibrium as reactants convert to products and vice versa. Thus the system is subject to minute fluctuations of concentrations of each species.

However, by the necessity of maintenance of the dynamic equilibrium the system always returns to the state of stable thermodynamic equilibrium. In a macroscopic sense then the system remains under the under equilibrium state described by eqn. 6.36b. It follows that in a reactive system at the state of chemical equilibrium the Gibbs free energy is minimum subject to the conditions of thermal and mechanical equilibrium.

The above considerations hold regardless of the number of reactants or the reactions occurring in the system. Since the reaction co-ordinate is the single parameter that relates the compositions of all the species, the variation of the total Gibbs free energy of the system as a function of the reaction co-ordinate may be shown schematically as in figure, here is the value of the reaction co-ordinate at equilibrium.

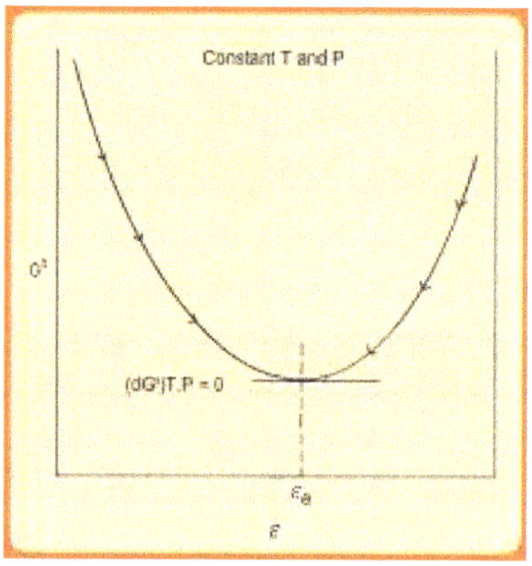

Variation of system Gibbs free energy with equilibrium conversion.

MASS FLUX

In physics and engineering, mass flux is the rate of mass flow per unit area, perfectly overlapping with the momentum density, the momentum per unit volume. The common symbols are j, J, q, Q, φ, or Φ, sometimes with subscript m to indicate mass is the flowing quantity. Its SI units are kg s^{-1} m^{-2}. Mass flux can also refer to an alternate form of flux in Fick's law that includes the molecular mass, or in Darcy's law that includes the mass density.

Mathematically, mass flux is defined as the limit:

$$j_m = \lim_{A \to 0} \frac{I_m}{A}$$

where,

$$I_m = \lim_{\Delta t \to 0} \frac{\Delta m}{\Delta t} = \frac{dm}{dt}$$

is the mass current (flow of mass *m* per unit time *t*) and *A* is the area through which the mass flows through.

For mass flux as a vector $\mathbf{j}_m$, the surface integral it over a surface *S*, followed by an integral over the time duration t_1 to t_2, gives the total amount of mass flowing through the surface in that time $(t_2 - t_1)$:

$$m = \int_{t_1}^{t_2} \iint_S \mathbf{j}_m \cdot \hat{\mathbf{n}} dA dt$$

The area required to calculate the flux is real or imaginary, flat or curved, either as a cross-sectional area or a surface.

For example, for substances passing through a filter or a membrane, the real surface is the (generally curved) surface area of the filter, macroscopically - ignoring the area spanned by the holes in the filter/membrane. The spaces would be cross-sectional areas. For liquids passing through a pipe, the area is the cross-section of the pipe, at the section considered.

The vector area is a combination of the magnitude of the area through which the mass passes through, *A*, and a unit vector normal to the area, $\hat{\mathbf{n}}$. The relation is $\mathbf{A} = A\hat{\mathbf{n}}$ If the mass flux j_m passes through the area at an angle θ to the area normal $\hat{\mathbf{n}}$, then

$$\mathbf{j}_m \cdot \hat{\mathbf{n}} = j_m \cos\theta$$

where • is the dot product of the unit vectors. This is, the component of mass flux passing through the surface (i.e. normal to it) is $j_m \cos\theta$, while the component of mass flux passing tangential to the area is $j_m \sin\theta$, but there is *no* mass flux actually passing *through* the area in the tangential direction. The *only* component of mass flux passing normal to the area is the cosine component.

Example:

Consider a pipe of flowing water. Suppose the pipe has a constant cross section and we consider a straight section of it (not at any bends/junctions), and the water is flowing steadily at a constant rate, under standard conditions. The area *A* is the cross-sectional area of the pipe. Suppose the pipe has radius $r = 2$ cm $= 2 \times 10^{-2}$ m. The area is then

$$A = \pi r^2$$

To calculate the mass flux j_m (magnitude), we also need the amount of mass of water transferred through the area and the time taken. Suppose a volume $V = 1.5$ L $= 1.5 \times 10^{-3}$ m^3 passes through in time $t = 2$ s. Assuming the density of water is $\rho = 1000$ kg m^{-3}, we have,

$$\Delta m = \rho \Delta V$$

$$m_2 - m_1 = \rho(V_2 - V_1)$$

$$m = \rho V$$

(since initial volume passing through the area was zero, final is V, so corresponding mass is m), so the mass flux is,

$$j_m = \frac{\Delta m}{A \Delta t} = \frac{\rho V}{\pi r^2 t}$$

substituting the numbers gives:

$$j_m = \frac{1000 \times (1.5 \times 10^{-3})}{\pi \times (2 \times 10^{-2})^2 \times 2} = \frac{3}{16\pi} \times 10^4$$

which is approximately 596.8 kg s^{-1} m^{-2}.

Equations for Fluids

Alternative Equation

Using the vector definition, mass flux is also equal to:

$$\mathbf{j}_\mathrm{m} = \rho \mathbf{u}$$

where,

- ρ = mass density.
- u = velocity field of mass elements flowing (i.e. at each point in space the velocity of an element of matter is some velocity vector u).

Sometimes this equation may be used to define $\mathbf{j}_\mathrm{m}$ as a vector.

Mass and Molar Fluxes for Composite Fluids

Mass Fluxes

In the case fluid is not pure, i.e. is a mixture of substances (technically contains a number of component substances), the mass fluxes must be considered separately for each component of the mixture.

When describing fluid flow (i.e. flow of matter), mass flux is appropriate. When describing particle transport (movement of a large number of particles), it is useful to use an analogous quantity, called the molar flux.

Using mass, the mass flux of component *i* is:

$$\mathbf{j}_{\mathrm{m},i} = \rho_i \mathbf{u}_i$$

The barycentric mass flux of component *i* is

$$\mathbf{j}_{\mathrm{m},i} = \rho\left(\mathbf{u}_i - \langle \mathbf{u} \rangle\right)$$

where, $\langle \mathbf{u} \rangle$ is the average mass velocity of all the components in the mixture, given by:

$$\langle \mathbf{u} \rangle = \frac{1}{\rho}\sum_i \rho_i \mathbf{u}_i = \frac{1}{\rho}\sum_i \mathbf{j}_{\mathrm{m},i}$$

where,

- ρ = mass density of the entire mixture,
- ρ_i = mass density of component *i*,
- u_i = velocity of component *i*.

The average is taken over the velocities of the components.

Molar Fluxes

If we replace density ρ by the number of moles *n*, we have the molar flux analogues.

The molar flux is the number of moles per unit time per unit area, generally:

$$\mathbf{j}_{\mathrm{n}} = c\mathbf{u}$$

So the molar flux of component *i* is (number of moles per unit time per unit area):

$$\mathbf{j}_{\mathrm{n},i} = c_i \mathbf{u}_i$$

and the barycentric molar flux of component *i* is,

$$\mathbf{j}_{\mathrm{n},i} = c\left(\mathbf{u}_i - \langle \mathbf{u} \rangle\right)$$

where $\langle \mathbf{u} \rangle$ this time is the average molar velocity of all the components in the mixture, given by:

$$\langle \mathbf{u} \rangle = \frac{1}{n}\sum_i c_i \mathbf{u}_i = \frac{1}{c}\sum_i \mathbf{j}_{\mathrm{n},i}$$

Usage

Mass flux appears in some equations in hydrodynamics, in particular the continuity equation:

$$\nabla \cdot \mathbf{j}_{\mathrm{m}} + \frac{\partial \rho}{\partial t} = 0$$

which is a statement of the mass conservation of fluid. In hydrodynamics, mass can only flow from one place to another.

Molar flux occurs in Fick's first law of diffusion:

$$\nabla \cdot \mathbf{j}_{\mathrm{n}} = -\nabla \cdot D \nabla n$$

where D is the diffusion coefficient.

FRICK'S LAW OF DIFFUSION

Fick's laws of diffusion describe diffusion and were derived by Adolf Fick in 1855. They can be used to solve for the diffusion coefficient, D. Fick's first law can be used to derive his second law which in turn is identical to the diffusion equation.

Fick's First Law

Fick's first law relates the diffusive flux to the concentration under the assumption of steady state. It postulates that the flux goes from regions of high concentration to regions of low concentration, with a magnitude that is proportional to the concentration gradient (spatial derivative), or in simplistic terms the concept that a solute will move from a region of high concentration to a region of low concentration across a concentration gradient. In one (spatial) dimension, the law can be written in various forms, where the most common form is in a molar basis:

$$J = -D \frac{d\varphi}{dx}$$

where,

- J is the diffusion flux, of which the dimension is amount of substance per unit area per unit time, so it is expressed in such units as mol m^{-2} s^{-1}. J measures the amount of substance that will flow through a unit area during a unit time interval.

- D is the diffusion coefficient or diffusivity. Its dimension is area per unit time, so typical units for expressing it would be m^2/s.

- φ (for ideal mixtures) is the concentration, of which the dimension is amount of substance per unit volume. It might be expressed in units of mol/m^3.
- x is position, the dimension of which is length. It might thus be expressed in the unit m.

D is proportional to the squared velocity of the diffusing particles, which depends on the temperature, viscosity of the fluid and the size of the particles according to the Stokes–Einstein relation. In dilute aqueous solutions the diffusion coefficients of most ions are similar and have values that at room temperature are in the range of (0.6–2)×10^{-9} m^2/s. For biological molecules the diffusion coefficients normally range from 10^{-11} to 10^{-10} m^2/s.

In two or more dimensions we must use ∇, the del or gradient operator, which generalises the first derivative, obtaining,

$$\mathbf{J} = -D\nabla\varphi$$

where, J denotes the diffusion flux vector.

The driving force for the one-dimensional diffusion is the quantity $-\frac{\partial\varphi}{\partial x}$, which for ideal mixtures is the concentration gradient.

Alternative Formulations of the First Law

Another form for the first law is to write it with the primary variable as mass fraction (y_i, given for example in kg/kg), for when the total concentration of the mixture is approximately constant, then the equation changes to:

$$J_i = -\rho D\nabla y_i$$

where,

- The index i denotes the ith species.
- J is still the same diffusion flux as in the common form, namely as amount of substance per unit area per unit time.
- ρ is the total concentration of the mixture, as in fluid density.
- y is the aforementioned mass fraction, which is dimensionless.

Note that the ρ is outside the gradient operator. This is because:

$$y_i = \frac{\rho_{si}}{\rho}$$

where, ρ_{si} is the concentration of the substance for the ith species.

Beyond this, in chemical systems other than ideal solutions or mixtures, the driving force for diffusion of each species is the gradient of chemical potential of this species. Then Fick's first law (one-dimensional case) can be written as:

$$J_i = -\frac{Dc_i}{RT}\frac{\partial \mu_i}{\partial x}$$

where,

- The index *i* denotes the *i*th species.
- *c* is the concentration (mol/m³).
- *R* is the universal gas constant (J/K/mol).
- *T* is the absolute temperature (K).
- μ is the chemical potential (J/mol).

Fick's Second Law

Fick's second law predicts how diffusion causes the concentration to change with respect to time. It is a partial differential equation which in one dimension reads:

$$\frac{\partial \varphi}{\partial t} = D\frac{\partial^2 \varphi}{\partial x^2}$$

where,

- φ is the concentration in dimensions of [(amount of substance) length^{-3}], example mol/m³; $\varphi = \varphi(x,t)$ is a function that depends on location *x* and time *t*.
- *t* is time, example s.
- *D* is the diffusion coefficient in dimensions of [length² time^{-1}], example m²/s.
- *x* is the position [length], example m.

In two or more dimensions we must use the Laplacian $\Delta = \nabla^2$, which generalises the second derivative, obtaining the equation,

$$\frac{\partial \varphi}{\partial t} = D\Delta\varphi$$

Derivation of Fick's Laws

Fick's second law can be derived from Fick's first law and the mass conservation in absence of any chemical reactions:

$$\frac{\partial \varphi}{\partial t} + \frac{\partial}{\partial x}J = 0 \Rightarrow \frac{\partial \varphi}{\partial t} - \frac{\partial}{\partial x}\left(D\frac{\partial}{\partial x}\varphi\right) = 0$$

Assuming the diffusion coefficient D to be a constant, one can exchange the orders of the differentiation and multiply by the constant:

$$\frac{\partial}{\partial x}\left(D\frac{\partial}{\partial x}\varphi\right)=D\frac{\partial}{\partial x}\frac{\partial}{\partial x}\varphi=D\frac{\partial^2\varphi}{\partial x^2}$$

and, thus, receive the form of the Fick's equations as was stated above.

For the case of diffusion in two or more dimensions Fick's second law becomes,

$$\frac{\partial\varphi}{\partial t}=D\nabla^2\varphi,$$

which is analogous to the heat equation.

If the diffusion coefficient is not a constant, but depends upon the coordinate or concentration, Fick's second law yields,

$$\frac{\partial\varphi}{\partial t}=\nabla\cdot(D\nabla\varphi).$$

An important example is the case where φ is at a steady state, i.e. the concentration does not change by time, so that the left part of the above equation is identically zero. In one dimension with constant D, the solution for the concentration will be a linear change of concentrations along x. In two or more dimensions we obtain,

$$\nabla^2\varphi=0$$

which is Laplace's equation, the solutions to which are referred to by mathematicians as harmonic functions.

Derivation

Fick's second law is a special case of the convection–diffusion equation in which there is no advective flux and no net volumetric source. It can be derived from the continuity equation:

$$\frac{\partial\varphi}{\partial t}+\nabla\cdot\mathbf{j}=R,$$

where j is the total flux and R is a net volumetric source for φ. The only source of flux in this situation is assumed to be diffusive flux:

$$\mathbf{j}_{\text{diffusion}}=-D\nabla\varphi$$

Plugging the definition of diffusive flux to the continuity equation and assuming there is no source ($R = 0$), we arrive at Fick's second law:

$$\frac{\partial \varphi}{\partial t} = D\frac{\partial^2 \varphi}{\partial x^2}$$

If flux were the result of both diffusive flux and advective flux, the convection–diffusion equation is the result.

Example Solution in one Dimension: Diffusion Length

A simple case of diffusion with time t in one dimension (taken as the x-axis) from a boundary located at position $x = 0$, where the concentration is maintained at a value n_0 is,

$$n(x,t) = n_0 \operatorname{erfc}\left(\frac{x}{2\sqrt{Dt}}\right).$$

where erfc is the complementary error function. This is the case when corrosive gases diffuse through the oxidative layer towards the metal surface (if we assume that concentration of gases in the environment is constant and the diffusion space – that is, the corrosion product layer – is *semi-infinite*, starting at 0 at the surface and spreading infinitely deep in the material). If, in its turn, the diffusion space is *infinite* (lasting both through the layer with $n(x,0) = 0, x > 0$ and that with $n(x,0) = n_0, x \leq 0$), then the solution is amended only with coefficient $\frac{1}{2}$ in front of n_0 (as the diffusion now occurs in both directions). This case is valid when some solution with concentration n_0 is put in contact with a layer of pure solvent. The length $2\sqrt{Dt}$ is called the *diffusion length* and provides a measure of how far the concentration has propagated in the x-direction by diffusion in time t.

As a quick approximation of the error function, the first 2 terms of the Taylor series can be used:

$$n(x,t) = n_0\left[1 - 2\left(\frac{x}{2\sqrt{Dt\pi}}\right)\right]$$

If D is time-dependent, the diffusion length becomes,

$$2\sqrt{\int_0^t D\tau d\tau}.$$

This idea is useful for estimating a diffusion length over a heating and cooling cycle, where D varies with temperature.

Generalizations

- In *non-homogeneous media*, the diffusion coefficient varies in space, $D = D(x)$. This dependence does not affect Fick's first law but the second law changes:

$$\frac{\partial\varphi(x,t)}{\partial t} = \nabla\cdot(D(x)\nabla\varphi(x,t)) = D(x)\Delta\varphi(x,t) + \sum_{i=1}^{3}\frac{\partial D(x)}{\partial x_i}\frac{\partial\varphi(x,t)}{\partial x_i}$$

- In *anisotropic media*, the diffusion coefficient depends on the direction. It is a symmetric tensor $D = D_{ij}$. Fick's first law changes to,

$$J = -D\nabla\varphi,$$

it is the product of a tensor and a vector:

$$J_i = -\sum_{j=1}^{3} D_{ij}\frac{\partial\varphi}{\partial x_j}.$$

For the diffusion equation this formula gives,

$$\frac{\partial\varphi(x,t)}{\partial t} = \nabla\cdot(D\nabla\varphi(x,t)) = \sum_{i=1}^{3}\sum_{j=1}^{3} D_{ij}\frac{\partial^2\varphi(x,t)}{\partial x_i\partial x_j}.$$

The symmetric matrix of diffusion coefficients D_{ij} should be positive definite. It is needed to make the right hand side operator elliptic.

- For *inhomogeneous anisotropic media* these two forms of the diffusion equation should be combined in,

$$\frac{\partial\varphi(x,t)}{\partial t} = \nabla\cdot(D(x)\nabla\varphi(x,t)) = \sum_{i,j=1}^{3}\left(D_{ij}(x)\frac{\partial^2\varphi(x,t)}{\partial x_i\partial x_j} + \frac{\partial D_{ij}(x)}{\partial x_i}\frac{\partial\varphi(x,t)}{\partial x_j}\right).$$

- The approach based on Einstein's mobility and Teorell formula gives the following generalization of Fick's equation for the *multicomponent diffusion* of the perfect components:

$$\frac{\partial\varphi_i}{\partial t} = \sum_j \nabla\cdot\left(D_{ij}\frac{\varphi_i}{\varphi_j}\nabla\varphi_j\right).$$

where, φ_i are concentrations of the components and D_{ij} is the matrix of coefficients. Here, indices *i* and *j* are related to the various components and not to the space coordinates.

The Chapman–Enskog formulae for diffusion in gases include exactly the same terms. These physical models of diffusion are different from the test models $\partial_t\varphi_i = \sum_j D_{ij}\Delta\varphi_j$

which are valid for very small deviations from the uniform equilibrium. Earlier, such terms were introduced in the Maxwell–Stefan diffusion equation.

For anisotropic multicomponent diffusion coefficients one needs a rank-four tensor, for example $D_{ij,\alpha\beta}$, where i, j refer to the components and $\alpha, \beta = 1, 2, 3$ correspond to the space coordinates.

Applications

Equations based on Fick's law have been commonly used to model transport processes in foods, neurons, biopolymers, pharmaceuticals, porous soils, population dynamics, nuclear materials, plasma physics, and semiconductor doping processes. Theory of all voltammetric methods is based on solutions of Fick's equation. Much experimental research in polymer science and food science has shown that a more general approach is required to describe transport of components in materials undergoing glass transition. In the vicinity of glass transition the flow behavior becomes "non-Fickian". It can be shown that the Fick's law can be obtained from the Maxwell–Stefan diffusion equations of multi-component mass transfer. The Fick's law is limiting case of the Maxwell–Stefan equations, when the mixture is extremely dilute and every chemical species is interacting only with the bulk mixture and not with other species. To account for the presence of multiple species in a non-dilute mixture, several variations of the Maxwell–Stefan equations are used.

Fick's Flow in Liquids

When two miscible liquids are brought into contact, and diffusion takes place, the macroscopic (or average) concentration evolves following Fick's law. On a mesoscopic scale, that is, between the macroscopic scale described by Fick's law and molecular scale, where molecular random walks take place, fluctuations cannot be neglected. Such situations can be successfully modeled with Landau-Lifshitz fluctuating hydrodynamics. In this theoretical framework, diffusion is due to fluctuations whose dimensions range from the molecular scale to the macroscopic scale.

In particular, fluctuating hydrodynamic equations include a Fick's flow term, with a given diffusion coefficient, along with hydrodynamics equations and stochastic terms describing fluctuations. When calculating the fluctuations with a perturbative approach, the zero order approximation is Fick's law. The first order gives the fluctuations, and it comes out that fluctuations contribute to diffusion. This represents somehow a tautology, since the phenomena described by a lower order approximation is the result of a higher approximation: this problem is solved only by renormalizing the fluctuating hydrodynamics equations.

Sorption Rate and Collision Frequency of Diluted Solute

The adsorption or absorption rate of a diluted solute to a surface or interface in a (gas or liquid) solution can be calculated using Fick's laws of diffusion, whose solution is typically a Gaussian function. Considering one dimension that is perpendicular to the surface, the probability of any given solute molecule in the solution hit the surface is the error function of its diffusive broadening over the time of interest. Thus integrate these error functions and integrate it with all solute molecules in the bulk gives the adsorption rate of the solute in unit S^{-1} to an area of interest:

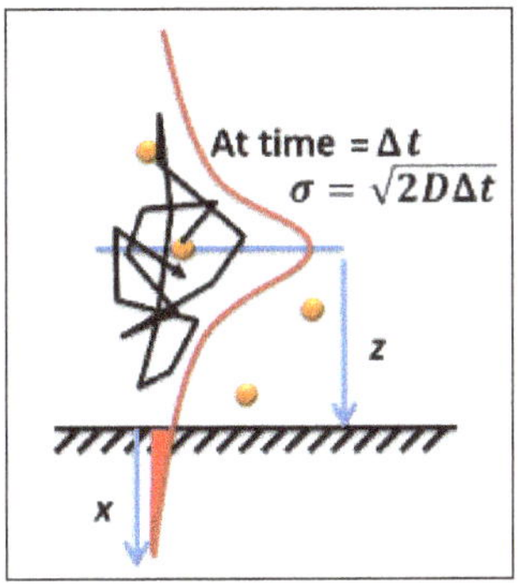

Scheme of molecular diffusion in the solution. Orange dots are solute molecules, solvent molecules are not drawn, black arrow is an example random walk trajectory, and the red curve is the diffusive Gaussian broadening probability function from the Fick's law of diffusion.

$$r = \frac{1}{t}\left(\int_{z=0}^{\infty} AC \int_{x=z}^{\infty} \frac{1}{\sqrt{4\pi Dt}} e^{-\frac{x^2}{4Dt}} dxdz \right) = \frac{1}{2} AC\sqrt{\pi D/t}$$

where, (all SI units)

- x is the distance of the probability function from the original location of a solute molecule (time t location references to its location at time 0, $\Delta t = t - 0$) (unit m).
- z is the original distance of the molecule from the surface (unit m).
- A is the surface area of the surface of interest (unit m^2).
- C is the number concentration of the molecule in the bulk solution (unit m^{-3}).
- D is the effective diffusion constant of the solute molecule measured at time resolution t.
- t is the time of interest (unit S).
- D is dependent on t, and the probability function is slightly non-Gaussian.

In the short time limit, in the order of the diffusion time a^2/D, where a is the particle radius, the diffusion is described by the Langevin equation. At a longer time, the Langevin equation merges into the Stokes-Einstein equation. The latter is appropriate for the condition of the diluted solution, where long-range diffusion is considered. According to the fluctuation-dissipation theorem based on the Langevin equation in the long-time limit and when the particle is significantly denser than the surrounding fluid, the time-dependent diffusion constant is:

$$D(t) = \mu k_B T(1 - e^{-t/(m\mu)})$$

where, (all SI units)

- k_B is Boltzmann's constant.
- T is the absolute temperature.
- μ is the mobility of the particle in the fluid or gas, which can be calculated using the Einstein relation (kinetic theory).
- m is the mass of the particle.
- t is time.

For a single molecule such as organic molecules or biomolecules (e.g proteins) in water, the exponential term is negligible due to the small product of $m\mu$ in the picosecond region.

When the area of interest is the size of a molecule (specifically, a *long cylindrical molecule* such as DNA), the adsorption rate equation represents the collision frequency of two molecules in a diluted solution, with one molecule a specific side and the other no steric dependence, i.e., a molecule (random orientation) hit one side of the other. The diffusion constant need to be updated to the relative diffusion constant between two diffusing molecules. This estimation is especially useful in studying the interaction between a small molecule and a larger molecule such as a protein. The effective diffusion constant is dominated by the smaller one whose diffusion constant can be used instead.

The above hitting rate equation is also useful to predict the kinetics of molecular self-assembly on a surface. Molecules are randomly oriented in the bulk solution. Assuming 1/6 of the molecules has the right orientation to the surface binding sites, i.e. 1/2 of the z-direction in x, y, z three dimensions, thus the concentration of interest is just 1/6 of the bulk concentration. Put this value into the equation one should be able to calculate the theoretical adsorption kinetic curve using the Langmuir adsorption model. In a more rigid picture, 1/6 can be replaced by the steric factor of the binding geometry.

Biological Perspective

The first law gives rise to the following formula:

$$\text{flux} = -P(c_2 - c_1)$$

in which,

- P is the permeability, an experimentally determined membrane "conductance" for a given gas at a given temperature.
- $c_2 - c_1$ is the difference in concentration of the gas across the membrane for the direction of flow (from c_1 to c_2).

Fick's first law is also important in radiation transfer equations. However, in this context it becomes inaccurate when the diffusion constant is low and the radiation becomes limited by the speed of light rather than by the resistance of the material the radiation is flowing through. In this situation, one can use a flux limiter.

The exchange rate of a gas across a fluid membrane can be determined by using this law together with Graham's law.

Under the condition of a diluted solution when diffusion takes control, the membrane permeability can be theoretically calculated for the solute (use with particular care because the equation is derived for dense solutes, while biological molecules are not denser than water):

$$P = \frac{1}{2} A_p \eta_{tm} \sqrt{\pi D t}$$

where,

- A_p is the total area of the pores on the membrane (unit m^2).
- η_{tm} transmembrane efficiency (unitless), which can be calculated from the stochastic theory of chromatography.
- D is the diffusion constant of the solute unit m^2s^{-1}.
- t is time unit s.
- c_2, c_1 concentration should use unit mol m^{-3}, so flux unit becomes mol s^{-1}.

Semiconductor Fabrication Applications

Integrated circuit fabrication technologies, model processes like CVD, thermal oxidation, wet oxidation, doping, etc. use diffusion equations obtained from Fick's law.

In certain cases, the solutions are obtained for boundary conditions such as constant source concentration diffusion, limited source concentration, or moving boundary diffusion (where junction depth keeps moving into the substrate).

DIFFUSION

Diffusion is the movement of a substance from an area of high concentration to an area of low concentration. It happens in liquids and gases because their particles move randomly from place to place.

Steady State Diffusion Through a Stagnant Gas Film

Assume steady state diffusion in the Z direction without any chemical reaction in a binary gaseous mixture of species A and B. For one dimensional diffusion of species A, the Equation of molar flux can be written as,

$$N_A = -CD_{AB}\frac{dy_A}{dZ} + y_A(N_A + N_B)$$

Separating the variables in Equation $N_A = -CD_{AB}\frac{dy_A}{dZ} + y_A(N_A + N_B)$, it can be expressed as,

$$\frac{-dy}{N_A - y_A(N_A + N_B)} = \frac{dZ}{CD_{AB}}$$

For the gaseous mixture, at constant pressure and temperature C and DAB are constant, independent of position and composition. Also all the molar fluxes are constant in Equation $\frac{-dy}{N_A - y_A(N_A + N_B)} = \frac{dZ}{CD_{AB}}$. Therefore the Equation $\frac{dy}{N_A - y_A(N_A + N_B)} \quad \frac{dZ}{CD_{AB}}$ can be integrated between two boundary conditions as follows:

at	$Z = Z_{1,}$	$y_A = y_{A1}$
at	$Z = Z_{2,}$	$y_A = y_{A1}$

where 1 indicates the start of the diffusion path and 2 indicates the end of the diffusion path. After integration with the above boundary conditions the Equation for diffusion for the said condition can be expressed as,

$$N_A = \frac{N_A}{N_A - y_B}\frac{CD_{AB}}{Z_2 - Z_1}\ln\left[\frac{\frac{N_A}{N_A - y_B} - y_{A2}}{\frac{N_A}{N_A - y_B} - y_{A1}}\right]$$

For steady state one dimensional diffusion of A through non-diffusing B, $N_B = 0$ and N_A = constant. Therefore $N_A/(N_A + N_B) = 1$.

Hence Equation $N_A = \frac{N_A}{N_A - y_B}\frac{CD_{AB}}{Z_2 - Z_1}\ln\left[\frac{\frac{N_A}{N_A - y_B} - y_{A2}}{\frac{N_A}{N_A - y_B} - y_{A1}}\right]$ becomes,

$$N_A = \frac{CD_{AB}}{Z_2 - Z_1}\ln\left[\frac{1 - y_{A2}}{y_{A1}}\right]$$

Since for an ideal gas $C = \frac{P}{RT}$ and for mixture of ideal gases $y_A = \frac{P_A}{P}$ =, the equation $N_A = \frac{CD_{AB}}{Z_2 - Z_1} \text{In}\left[\frac{1-y_{A2}}{y_{A1}}\right]$ can be expressed in terms of partial pressures as,

$$N_A = \frac{PD_{AB}}{(Z_2 - Z_1)RT} \text{In}\left[\frac{P-P_{A2}}{P-P_{A1}}\right]$$

Where P is the total pressure and p_{A1} and p_{A2} are the partial pressures of A at point 1 and 2 respectively. For diffusion under turbulent conditions, the flux is usually calculated based on linear driving force. For this purpose the equation $N_A = \frac{N_A}{N_A - y_B} \frac{CD_{AB}}{Z_2 - Z_1} \text{In}\left[\frac{\frac{N_A}{N_A - y_B} - y_{A2}}{\frac{N_A}{N_A - y_B} - y_{A1}}\right]$ can be manipulated to rewrite it in terms of a linear driving force. Since for the binary gas mixture of total pressure P, P- A_2 =B_2;

$P - p = p_{A1} = P_{B1}$ $P_{A1} - P_{A2} = P_{B2} - P_{B1}$. Then the equation $N_A = \frac{PD_{AB}}{(Z_2 - Z_1)RT} \text{In}\left[\frac{P-P_{A2}}{P-P_{A1}}\right]$ can be written as,

$$N_A = \frac{PD_{AB}}{(Z_2 - Z_1)RT} \text{In}\left[\frac{P_{A1} - P_{A2}}{P_{B2} - P_{B1}}\right] \text{In}\left[\frac{P_{A2}}{P_{B1}}\right]$$

Or

$$N_A = \frac{PD_{AB}}{(Z_2 - Z_1)RTp_{B,M}}(p_{A1} - p_{A2})$$

Where $p_{B,M}$ is called logarithmic mean partial pressure of species B which is defined as

$$p_{B,M} = \frac{P_{B2} - P_{B1}}{\text{In}\left(\frac{P_{B2}}{P_{B1}}\right)}$$

A schematic concentration profile for diffusion A through stagnant B is shown in figure. The component A diffuses by concentration gradient, $-\frac{dy_A}{dZ}$. Here flux is inversely proportional to the distance through which diffusion occurs and the concentration of the stagnant gas ($p_{B,M}$) because with increase in Z and $p_{B,M}$, resistance increases and flux decreases.

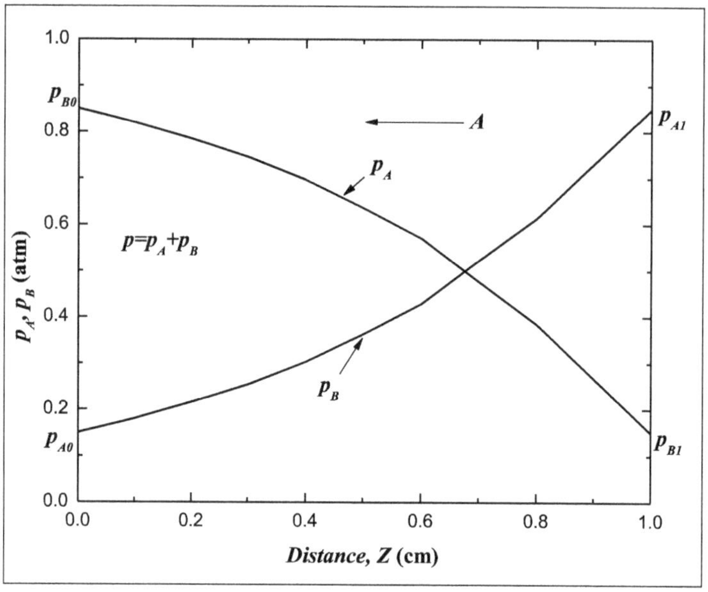

Partial pressure distribution of A in non-diffusing B.

Steady State Equimolar Counter Diffusion

This is the case for the diffusion of two ideal gases, where an equal number of moles of the gases diffusing counter-current to each other. In this case $N_B = -N_A$ = constant and $N_A + N_B = 0$. The molar flux Equation ($N_A = -CD_{AB}\frac{dy_A}{dZ} + y_A(N_A + N_B)$) at steady state can then be written as,

$$N_A - \frac{d_{AB}p}{RT}\frac{dy_A}{dZ}$$

Integrating the Equation $N_A - \frac{d_{AB}p}{RT}\frac{dy_A}{dZ}$ with the boundary conditions: at $Z = Z_1$, $y_A = y_{A1}$; at $Z = Z_2$ $y_A = y_{A2}$, the Equation of molar diffusion for steady-state equimolar counter diffusion can be represented as,

$$N_A - \frac{d_{AB}p}{RT(Z_2 - Z_1)}(y_{A1} - y_{A2})$$

$$= \frac{D_{AB}}{RT(Z_2 - Z_1)}(P_{A1} - P_{A2})$$

It may be noted here also that molar latent heats of vaporization of A and B are equal. So, $\Delta H_A^V = \Delta H_B^V$, where, $\Delta H_A^V = \Delta H_B^V$ are molar latent heats of vaporization of A and B, respectively. The concentration profile in terms of partial pressure is shown in figure.

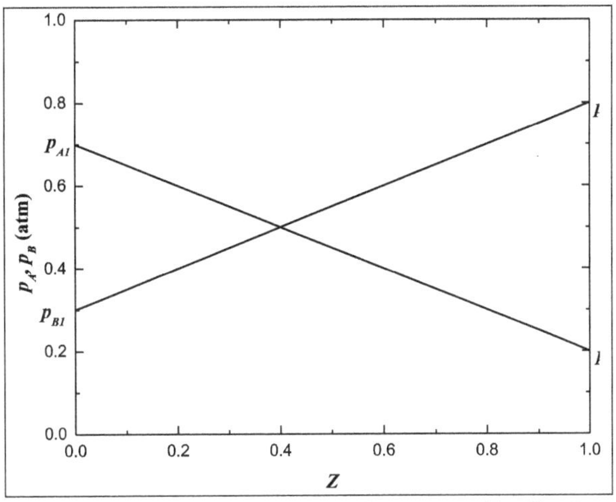

Equimolar counter diffusion of A and B: Partial pressure distribution with position.

Non-equimolar Counter Diffusion

In some practical cases, A and B molecules diffuse in opposite directions at different molar velocities. Let carbon monoxide is generated from the reaction between hot char and oxygen. The stoichiometry is as follows:

$$2C + O_2\,(A) \leftrightarrow 2CO(B)$$

When one mole oxygen molecule diffuses towards char, two moles carbon monoxide molecules diffuse in opposite direction. Here, NA = –NB / 2 and molar latent heats of vaporization are not equal. Hence,

$$N_A\,\Delta H_A^V = -N_B \Delta H_B^V.$$

References

- Bian, xin; kim, changho; karniadakis, george em (14 august 2016). "111 years of brownian motion". Soft matter. 12 (30): 6331–6346. Bibcode:2016smat...12.6331b. Doi:10.1039/c6sm01153e. Pmc 5476231. Pmid 27396746
- What-is-mass-transfer, multiphysics: comsol.co.in, Retrieved 30 January, 2019
- Fluid mechanics, m. Potter, d.c. wiggart, schuam's outlines, mcgraw hill (usa), 2008, isbn 978-0-07-148781-8
- gorban, a. N.; sargsyan, h. P.; wahab, h. A. (2011). "quasichemical models of multicomponent nonlinear diffusion". Mathematical modelling of natural phenomena. 6 (5): 184–262. Arxiv:1012.2908. Doi:10.1051/mmnp/20116509
- Conlisk, a. Terrence (2013). Essentials of micro- and nanofluidics: with applications to the biological and chemical sciences. Cambridge university press. P. 43. Isbn 9780521881685
- Brogioli, d.; vailati, a. (2001). "diffusive mass transfer by nonequilibrium fluctuations: fick's law revisited". Phys. Rev. E. 63 (1–4): 012105. Arxiv:cond-mat/0006163. Bibcode:2001phrve..63a2105b. Doi:10.1103/physreve.63.012105. Pmid 11304296

Permissions

All chapters in this book are published with permission under the Creative Commons Attribution Share Alike License or equivalent. Every chapter published in this book has been scrutinized by our experts. Their significance has been extensively debated. The topics covered herein carry significant information for a comprehensive understanding. They may even be implemented as practical applications or may be referred to as a beginning point for further studies.

We would like to thank the editorial team for lending their expertise to make the book truly unique. They have played a crucial role in the development of this book. Without their invaluable contributions this book wouldn't have been possible. They have made vital efforts to compile up to date information on the varied aspects of this subject to make this book a valuable addition to the collection of many professionals and students.

This book was conceptualized with the vision of imparting up-to-date and integrated information in this field. To ensure the same, a matchless editorial board was set up. Every individual on the board went through rigorous rounds of assessment to prove their worth. After which they invested a large part of their time researching and compiling the most relevant data for our readers.

The editorial board has been involved in producing this book since its inception. They have spent rigorous hours researching and exploring the diverse topics which have resulted in the successful publishing of this book. They have passed on their knowledge of decades through this book. To expedite this challenging task, the publisher supported the team at every step. A small team of assistant editors was also appointed to further simplify the editing procedure and attain best results for the readers.

Apart from the editorial board, the designing team has also invested a significant amount of their time in understanding the subject and creating the most relevant covers. They scrutinized every image to scout for the most suitable representation of the subject and create an appropriate cover for the book.

The publishing team has been an ardent support to the editorial, designing and production team. Their endless efforts to recruit the best for this project, has resulted in the accomplishment of this book. They are a veteran in the field of academics and their pool of knowledge is as vast as their experience in printing. Their expertise and guidance has proved useful at every step. Their uncompromising quality standards have made this book an exceptional effort. Their encouragement from time to time has been an inspiration for everyone.

The publisher and the editorial board hope that this book will prove to be a valuable piece of knowledge for students, practitioners and scholars across the globe.

Index

www.ingramcontent.com/pod-product-compliance
Lightning Source LLC
Chambersburg PA
CBHW062002190326
41458CB00009B/2941
* 9 7 8 1 6 3 2 3 8 8 0 7 0 *